Certifying Surgical First Assistant Review Workbook

1st Edition

10 Actual case studies
1,000 Practice questions

Lonnie Bargo CST/CFA

Certifying Surgical First Assistant Review Workbook Review Workbook 1st Edition
Published through Lulu Enterprises, Inc.

ISBN: 978-0-615-26489-9

Certifying Surgical First Assistant Review

I would like to dedicate this book to my family who have stood behind me, my friends that believed in me, and my mom and dad who still say that I can do anything.

Contents

Bibliography

Seeley, Stephens, and Tate (2006) *Anatomy & Physiology*,
Boston, McGraw Hill

Price, Frey, and Junge (2004) *Surgical Technology for the Surgical Technologist*,
United States, Thomson Delmar Learning

Way (1994) Current Stephens, *Surgical Diagnosis & Treatment,*
Norwalk, Connecticut, Appleton & Lange

Edmonson, Crenshaw (1980) *Campbell's Operative Orthopaedics Vol 1 and Vol 2,*
ST. Louis, The C.V. Mosby Company

Fuller, (2005) *Surgical Technology Principals and Practice,*
United States, Evolve, Elsevier Sanders

Thomas Dolan, Teaching and Academic Support Center, Medical illustrator and multimedia developer, University of Kentucky

Exam Providers

Association of Surgical Technologists
6 West Dry Creek Circle, Suite 200
Littleton CO 80120
http://www.ast.org/
800-637-7433 toll free
(303) 694-9130 voice
(303) 694-9169 fax

National Surgical Assistant Association
2615 Amesbury Road
Winston Salem, NC 27103
http://www.nsaa.net/index.shtml
Voice (336)768-4443
Fax (336)768-4445
Toll free1-888-633-0479

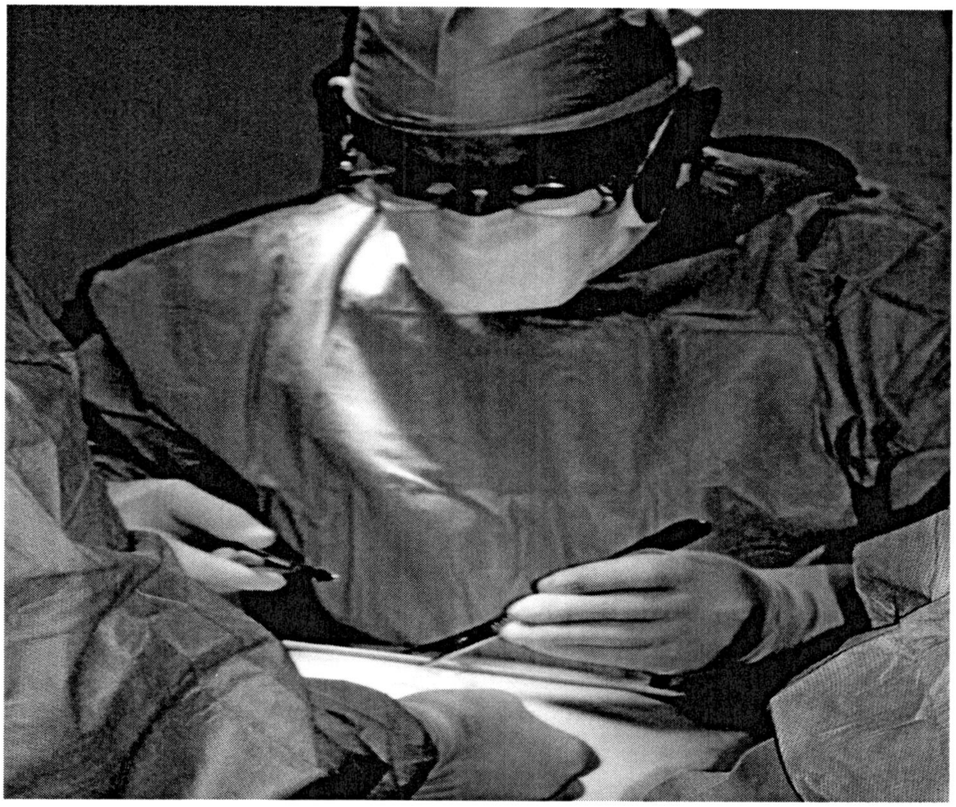

Preface

This textbook offers a variety of learning tools to the first assistant in training. It will better prepare you for your clinical rotation as well as the certifying exam. Each chapter in this book is setup the same way. It starts with an overview of a particular surgery and explains from start to finish from the first assistant's point of view. Next are pages for you the student to use for your version of this or a similar surgery. Each chapter has 100 questions asked in two formats, multiple choice and true or false. The questions are setup so when a chapter is finished you will have better understanding of the anatomy and reasons things are done this way. There are exam sheets and the correct answers at the end of each chapter.

To be a strong first assistant you must learn as much as possible on all surgical specialties such as; tests ran, positioning, drugs and their interactions, wound healing, and about skin integrity. This lists only part of the vastness of information that one must know before attempting to be left in charge of the patient when the surgeon leaves the room. While in clinical rotation you should volunteer when possible to help with all aspects of the surgery to gain this needed knowledge. I hope this book helps you in surgery as well as give you an understanding of what to expect when taking one of the certifying exams.

CHAPTER 1

ACROMIOPLASTY CASE STUDY

Acromioplasty

This afternoon the first case on the board for us is an Acromioplasty. I will need to stop by x-ray and pick up the patients folder which contains x-rays and cat scans. On my way back I will check the chart for any known allergies, previous surgeries, or any other abnormalities. I will also speak to the patient, identify myself and ask if there are any questions that she might have regarding the surgery. I will make the team aware of anything extra that they may need. When I get to the room, I will then ensure that the room is ready for the patient: x-rays hung, proper drugs available, the case is open and all instrumentation is on hand and working properly, that anesthesia has spoken to the patient and we are ready to bring the patient to the room. The reason we are doing this surgery today is that the patient has impingement syndrome that is limiting her range of motion and causing pain.

While the anesthesiologist is sedating the patient, I will assist with intubations by applying cricoid pressure (Selleck's procedure) if needed. With the patient asleep I can now place the Foley, and since this is a female I will only need to insert the catheter about four inches. Now we prepare to place the patient into semi-Fowler's position. With the help of all team members we will slowly move our patient to the edge of the table, place a roll under the affected shoulder, with the arm free, while anesthesia protects the head and airway. Then I will place the unaffected arm on a padded arm board out from the side. The head will be placed in a headrest and taped down securely as to not enter the field.

Now I will need to place the bovie pad over a fatty location but still close to incision site. I will carefully place warm blankets over the arms and legs of my patient to prevent any hypothermia. The table strap is placed last to prevent the patient from a fall when the table is repositioned. With the site properly marked I can beep the surgeon and go scrub while the circulator does the prep. The Surgical technologist and I will drape the patient and the arm, as well as hook up bovie, suction, and power source while the surgeon is scrubbing. A surgical timeout is done as to recognize the proper patient with their surgery and we can begin.

The incision we are going to use is oblique and anterolateral to the Acromion. We only extend the incision as to get the proper exposer. I will use skin rakes, as the super fiscal layers of tissue need to be exposed.

With use of the bovie the fascia is freed from the deltoid muscle and the deltoid is split with the bovie and/or scissors. I will then place two gelpi retractors into the wound to get better exposure of the acromion. The deltoid will then be dissected off of the anterior acromion and the coracoacromial ligament is dissected off the underside of the acromion.

The technique used is called the Rockwood where we resect the distal end of the acromion with a Stryker sagittal saw. We will then use a file and small ronjeur to smooth any sharp edges that we might have. The resulting cut will allow the shoulder to move freely without impinging any tissue. Last we are going to irrigate with bacitracin in our saline and prepare to close. For my closure I will use a number 2 Vicryl for strength and interrupt those, next a 2-0 Vicryl running stitch, and finishing the closure with a 4-0 Monocryl sub cue. After an injection of 10cc of Marcaine I will apply steri strips and mastasol. The dressing will be antibiotic ointment, adptic, fluffs, and medipore tape.

The drapes are then removed and the table will return to level so that a sling and swath can be placed and the patient extubated. Care will be taken as we proceed to move the patient to her stretcher and transported to recovery. On my follow up I will advise the patient on dressing changes, showers, and any mobilization of her shoulder. The expected outcome will be to return to normal activities within six weeks.

Case study

In this section you will need to describe an Acromioplasty in detail from your perspective (one that you have done in the past). List any differences in positioning, drugs used, incision, etc.

1. The reason for doing an acromioplasty is
   A. Impingement syndrome
   B. Humeral fracture
   C. Capsule displacement
   D. Bursitis

2. The procedure of applying cricoid pressure is called
   A. Bannon procedure
   B. Wolton
   C. Selleck's procedure
   D. Robinson procedure

3. Beach chair position is same as
   A. Supine
   B. Semi-Fowler's
   C. Prone
   D. Simm's

4. Core body temperature lower than 35^0 C will be
   A. Hyperthermia
   B. Hypovolemic
   C. Hypo anesthesia
   D. Hypothermia

5. Incision most used for acromioplasty is a/an

   A. Kocker

   B. Rockwood

   C. Supralateral

   D. Oblique anterolateral

6. Blunt and sharp dissection is used on which muscle during this procedure?

   A. Biceps brachii

   B. Sternocleidomastoid

   C. Pectoralis major

   D. Deltoid

7. The Deltoid muscle is dissected from the

   A. Anterior acromion

   B. Superior acromion

   C. Scapula

   D. Clavicle

8. Which ligament is dissected off of the under side of the acromion?

   A. Coracoid process

   B. Coracoacromial

   C. Pectoral

   D. Scapular

9. A common technique for this procedure

A. Rockwood

B. Kocker

C. Penn

D. Bannon

10. Common antibiotic irrigation for any exposed or cut bone is

A. Gentamicin

B. Vancomycin

C. Bacitracin

D. Ancef

11. A long curved bone, prop for the shoulder, articulates medially with sternum, and laterally with acromion

A. Scapula

B. Acromion

C. Sternum

D. Clavicle

12. A flat, triangular bone, posterior part of the shoulder girdle is

A. Scapula

B. Acromion

C. Sternum

D. Clavicle

13. Glenohumeral, coracoclavicular, and acromioclavicular joints. These three articulations make up

A. Sternum

B. Acromion

C. Pelvic girdle

D. Shoulder girdle

14. The acromioclavicular joint has

A. Full motion

B. Limited motion

C. No motion

D. Multidirectional motion

15. The glenohumeral joint has

A. Full motion

B. Limited motion

C. No motion

D. Multidirectional motion

16. The coracoclavicular joint has

A. Full motion

B. Limited motion

C. No motion

D. Multidirectional motion

17. The longest and largest bone of the upper extremity is the

A. Humerus

B. Radius

C. Ulna

D. Clavicle

18. The site of most humeral fractures is

A. The olecranon

B. Above the tuberosities

C. Below the tuberosities

D. The surgical neck

19. The cavity on the lateral side of the clavicle for the humerus to rest in is

A. Clavicle

B. Sternum

C. Glenoid

D. Acromion

20. The scapula is attached to the _________by muscle

A. Olecranon

B. Humerus

C. Glenoid

D. Thorax

21. What is meant by the phrase appendicular skeleton?

    A. Upper and lower appendages

    B. Skull

    C. Thorax

    D. All bones

22. Trochlea and the capitulum are

    A. Two humeral muscles

    B. Two ligaments

    C. Two humeral condyles

    D. Two bones

23. The olecranon fossa of the humerus receives the olecranon process of the

    A. Radius

    B. Ulna

    C. Glenoid

    D. Deltoid tuberosity

24. Muscles connecting the arms to the axial skeleton attach to the

    A. Thorax

    B. Scapula and clavicle

    C. Glenoid

    D. Deltoid tuberosity

25. The deltoid muscle connects to the humerus at the

    A. Thorax

    B. Scapula and clavicle

    C. Glenoid

    D. Deltoid tuberosity

26. The appendicular skeleton contains how many bones

    A. 33

    B. 18

    C. 151

    D. 126

27. The deltoid tuberosity is located where on the humerus?

    A. On the humeral head

    B. Distal shaft

    C. Mid shaft

    D. Proximal shaft

28. The coronoid fossa on the humerus is located where?

    A. On the humeral head

    B. Distal shaft

    C. Mid shaft

    D. Proximal shaft

29. Which joint has the widest range of motion?

    A. Ball and socket

    B. Saddle

    C. Gliding

    D. Pivot

30. Which joint moves in a single plane?

    A. Ball and socket

    B. Hinge

    C. Saddle

    D. Condyloid

31. The muscle in the neck that flexes the head is the

    A. Trapezius

    B. Pectoralis

    C. Deltoid

    D. Sternocleidomastoid

32. The muscle in the shoulder that abducts the upper arm is the

    A. Trapezius

    B. Pectoralis

    C. Deltoid

    D. Sternocleidomastoid

33. The muscle in the shoulder that adducts the upper arm is the

    A. Trapezius

    B. Deltoid

    C. Pectoralis major

    D. Brachii

34. The muscle in the back that extends the upper arm is the

    A. Trapezius

    B. Deltoid

    C. Pectoralis major

    D. Brachii

35. The muscle in the shoulder that abducts the shoulder

    A. Serratus anterior

    B. Deltoid

    C. Pectoralis major

    D. Brachii

36. This large triangular muscle originates on bone and cartilages of the thorax and inserts on the humerus

    A. Trapezius

    B. Pectoralis major

    C. Deltoid

    D. Sternocleidomastoid

37. This muscle is referred to as the shoulder pad

    A. Trapezius

    B. Pectoralis

    C. Deltoid

    D. Sternocleidomastoid

38. This small triangular muscle lies under the pectoralis major

    A. Pectoralis minor

    B. Deltoid

    C. Brachii

    D. Serratus anterior

39. This muscle with its two heads has it origin on the scapula and inserts into the radius

    A. Pectoralis minor

    B. Deltoid

    C. Brachii

    D. Biceps brachii

40. The muscle that reaches from the scapula to the humerus then flexes and abducts the upper arm is the

    A. Pectoralis minor

    B. Deltoid

    C. Corocobrachialis

    D. Biceps brachii

41. The three nerves that make up the brachial plexus are

   A. Axillary, Median, Ulnar

   B. Radial, Median, Ulnar

   C. Radial, Median, Axillary

   D. Axillary, Radial, Ulnar

42. This large artery is a continuation of the subclavian artery

   A. Thoracic

   B. Ulnar

   C. Radial

   D. Axillary

43. This nerve may be damaged by dislocation of the shoulder joint

   A. Thoracic

   B. Ulnar

   C. Radial

   D. Axillary

44. The brachial artery continues off of the

   A. Thoracic

   B. Ulnar

   C. Radial

   D. Axillary

45. Which artery arises off the axillary to supply the pectoralis minor?

A. Thoracoacromial

B. Ulnar

C. Radial

D. Axillary

46. On the left side of the body which artery rises out of the aorta along with the subclavian?

A. Vertebral

B. Lateral thoracic

C. Carotid

D. Internal thoracic

47. The subclavian artery passes under the clavicle where?

A. Mid shaft

B. Distally

C. Proximally

D. It doesn't

48. The axillary vein continues to the

A. Subclavian

B. Ulnar

C. Radial

D. Thoracic

49. Wounds in the axilla often involve the axillary vein because of
    A. Size
    B. Location
    C. Frailty
    D. A and B

50. The axillary vein passes under the clavicle
    A. Mid shaft
    B. Distally
    C. Proximally
    D. It doesn't

51. Usually the reason for doing an acromioplasty is impingement syndrome
    A. True
    B. False

52. Selleck's procedure is the act of applying cricoid pressure
    A. True
    B. False

53. Beach chair position is Semi-Fowler's
    A. True
    B. False

54. Hyperthermia is core body temperature of less than 35^0C
    A. True
    B. False

55. A Kocker incision is most common for an acromioplasty

    A. True

    B. False

56. In an acromioplasty it's best to use sharp and blunt dissection on the deltoid

    A. True

    B. False

57. The deltoid muscle is dissected off the anterior acromion during shoulder repair

    A. True

    B. False

58. The coracoacromial is dissected off the acromion during shoulder repair

    A. True

    B. False

59. The Rockwood is a common type of shoulder repair

    A. True

    B. False

60. Vancomycin is common antibiotic for irrigation on bone cases

    A. True

    B. False

61. The clavicle is a long curved bone, prop for the shoulder, articulates medially with sternum, and laterally with acromion

A. True

B. False

62. The clavicle is a flat, triangular bone, posterior part of the shoulder girdle

A. True

B. False

63. The acromioclavicular joint has multi -directional motion

A. True

B. False

64. The three articulations that make up the pelvic girdle are glenohumeral, coracoclavicular, and acromioclavicular joints

A. True

B. False

65. The glenohumeral joint has multi-directional motion

A. True

B. False

66. The coracoclavicular joint has limited motion

A. True

B. False

67. The humerus is the longest and largest bone of the extremity

    A. True

    B. False

68. The humeral neck is the site of most fractures

    A. True

    B. False

69. The glenoid is the cavity on the lateral side in which the humerus rests in

    A. True

    B. False

70. The scapula and the thorax are connected by muscle

    A. True

    B. False

71. All the bones of the skeletal system are considered the appendicular skeleton

    A. True

    B. False

72. Trochlea and the capitulum are two humeral condyles

    A. True

    B. False

73. The olecranon fossa of the humerus receives the olecranon process of the ulna

    A. True

    B. False

74. Muscles connecting the arms to the axial skeleton attach to the scapula and clavicle

A. True

B. False

75. The deltoid muscle connects to the humerus at the olecranon tuberosity

A. True

B. False

76. The appendicular skeleton contains 126 bones

A. True

B. False

77. The deltoid tuberosity is located mid-shaft on the humerus

A. True

B. False

78. The coronoid fossa on the humerus is located on the distal shaft

A. True

B. False

79. A ball and socket joint has the widest range of motion

A. True

B. False

80. A ball and socket joint moves in a single plane

A. True

B. False

81. The deltoid muscle flexes the head

A. True

B. False

82. The deltoid in the shoulder abducts the upper arm

A. True

B. False

83. The deltoid in the shoulder adducts the upper arm

A. True

B. False

84. The trapezius is the muscle in the back that abducts the shoulder

A. True

B. False

85. The serratus anterior is one of the muscles located in the shoulder that abducts the shoulder

A. True

B. False

86. The trapezius is a large triangular muscle that originates on bone and cartilages of the thorax and inserts on the humerus

A. True

B. False

87. The deltoid is referred to as the shoulder pad

A. True

B. False

88. The pectoralis minor is a small triangular muscle that lies under the pectoralis major

A. True

B. False

89. Biceps brachii is a muscle with two heads that have their origin on the scapula and inserts into the radius

A. True

B. False

90. The muscle that reaches from the scapula to the humerus then flexes and abducts the upper arm is the Corocobrachialis

A. True

B. False

91. Axillary, Median, and Ulnar are the three nerves that makeup the brachial plexus

A. True

B. False

92. The thoracic artery is a large artery that communicates with the subclavian artery

A. True

B. False

93. The axillary nerve can be damaged by dislocation of the shoulder joint

A. True

B. False

94. The brachial artery continues off of the axillary artery

A. True

B. False

95 The thorocoacromial artery arises off the axillary to supply the pectoralis minor

A. True

B. False

96. On the left side of the body the carotid artery rises out of the aorta along with the subclavian

A. True

B. False

97. The subclavian artery passes over the clavicle

A. True

B. False

98. The axillary vein is a continuation of the subclavian vein

    A. True

    B. False

99. Size and location are contributing factors in the axillary vein often being damaged when the axilla is wounded

    A. True

    B. False

100. The axillary vein passes under the clavicle near mid-shaft

    A. True

    B. False

1 A ❑ B ❑ C ❑ D ❑
2 A ❑ B ❑ C ❑ D ❑
3 A ❑ B ❑ C ❑ D ❑
4 A ❑ B ❑ C ❑ D ❑
5 A ❑ B ❑ C ❑ D ❑
6 A ❑ B ❑ C ❑ D ❑
7 A ❑ B ❑ C ❑ D ❑
8 A ❑ B ❑ C ❑ D ❑
9 A ❑ B ❑ C ❑ D ❑
10 A ❑ B ❑ C ❑ D ❑
11 A ❑ B ❑ C ❑ D ❑
12 A ❑ B ❑ C ❑ D ❑
13 A ❑ B ❑ C ❑ D ❑
14 A ❑ B ❑ C ❑ D ❑
15 A ❑ B ❑ C ❑ D ❑
16 A ❑ B ❑ C ❑ D ❑
17 A ❑ B ❑ C ❑ D ❑
18 A ❑ B ❑ C ❑ D ❑
19 A ❑ B ❑ C ❑ D ❑
20 A ❑ B ❑ C ❑ D ❑
21 A ❑ B ❑ C ❑ D ❑
22 A ❑ B ❑ C ❑ D ❑
23 A ❑ B ❑ C ❑ D ❑
24 A ❑ B ❑ C ❑ D ❑
25 A ❑ B ❑ C ❑ D ❑
26 A ❑ B ❑ C ❑ D ❑
27 A ❑ B ❑ C ❑ D ❑
28 A ❑ B ❑ C ❑ D ❑
29 A ❑ B ❑ C ❑ D ❑
30 A ❑ B ❑ C ❑ D ❑
31 A ❑ B ❑ C ❑ D ❑
32 A ❑ B ❑ C ❑ D ❑
33 A ❑ B ❑ C ❑ D ❑
34 A ❑ B ❑ C ❑ D ❑
35 A ❑ B ❑ C ❑ D ❑
36 A ❑ B ❑ C ❑ D ❑
37 A ❑ B ❑ C ❑ D ❑
38 A ❑ B ❑ C ❑ D ❑
39 A ❑ B ❑ C ❑ D ❑
40 A ❑ B ❑ C ❑ D ❑
41 A ❑ B ❑ C ❑ D ❑
42 A ❑ B ❑ C ❑ D ❑
43 A ❑ B ❑ C ❑ D ❑
44 A ❑ B ❑ C ❑ D ❑
45 A ❑ B ❑ C ❑ D ❑
46 A ❑ B ❑ C ❑ D ❑
47 A ❑ B ❑ C ❑ D ❑
48 A ❑ B ❑ C ❑ D ❑
49 A ❑ B ❑ C ❑ D ❑
50 A ❑ B ❑ C ❑ D ❑
51 A ❑ B ❑ C ❑ D ❑
52 A ❑ B ❑ C ❑ D ❑
53 A ❑ B ❑ C ❑ D ❑
54 A ❑ B ❑ C ❑ D ❑
55 A ❑ B ❑ C ❑ D ❑
56 A ❑ B ❑ C ❑ D ❑
57 A ❑ B ❑ C ❑ D ❑
58 A ❑ B ❑ C ❑ D ❑
59 A ❑ B ❑ C ❑ D ❑
60 A ❑ B ❑ C ❑ D ❑
61 A ❑ B ❑ C ❑ D ❑
62 A ❑ B ❑ C ❑ D ❑
63 A ❑ B ❑ C ❑ D ❑
64 A ❑ B ❑ C ❑ D ❑
65 A ❑ B ❑ C ❑ D ❑
66 A ❑ B ❑ C ❑ D ❑
67 A ❑ B ❑ C ❑ D ❑
68 A ❑ B ❑ C ❑ D ❑
69 A ❑ B ❑ C ❑ D ❑
70 A ❑ B ❑ C ❑ D ❑
71 A ❑ B ❑ C ❑ D ❑
72 A ❑ B ❑ C ❑ D ❑
73 A ❑ B ❑ C ❑ D ❑
74 A ❑ B ❑ C ❑ D ❑
75 A ❑ B ❑ C ❑ D ❑
76 A ❑ B ❑ C ❑ D ❑
77 A ❑ B ❑ C ❑ D ❑
78 A ❑ B ❑ C ❑ D ❑
79 A ❑ B ❑ C ❑ D ❑
80 A ❑ B ❑ C ❑ D ❑
81 A ❑ B ❑ C ❑ D ❑
82 A ❑ B ❑ C ❑ D ❑
83 A ❑ B ❑ C ❑ D ❑
84 A ❑ B ❑ C ❑ D ❑
85 A ❑ B ❑ C ❑ D ❑
86 A ❑ B ❑ C ❑ D ❑
87 A ❑ B ❑ C ❑ D ❑
88 A ❑ B ❑ C ❑ D ❑
89 A ❑ B ❑ C ❑ D ❑
90 A ❑ B ❑ C ❑ D ❑
91 A ❑ B ❑ C ❑ D ❑
92 A ❑ B ❑ C ❑ D ❑
93 A ❑ B ❑ C ❑ D ❑
94 A ❑ B ❑ C ❑ D ❑
95 A ❑ B ❑ C ❑ D ❑
96 A ❑ B ❑ C ❑ D ❑
97 A ❑ B ❑ C ❑ D ❑
98 A ❑ B ❑ C ❑ D ❑
99 A ❑ B ❑ C ❑ D ❑
100 A ❑ B ❑ C ❑ D ❑

CH1 Acromioplasty

Multiple Choice

1.A	11.D	21.A	31D.	41.B
2.C	12.A	22.C	32.C	42.D
3.B	13.D	23.B	33.C	43.D
4.D	14.B	24.B	34.A	44.D
5.D	15.D	25.D	35.A	45.A
6.D	16.B	26.D	36.B	46.C
7.A	17.A	27.C	37.C	47.A
8.B	18.D	28.B	38.A	48.A
9.A	19.C	29.A	39.D	49.D
10.C	20.D	30.B	40.C	50.A

True / False

51. True	61. True	71. Appendages	81.Sternocliedo-mastoid	91. Rad, med, ,ul
52. True	62. Scapula	72. True	82. True	92. Axillary
53. True	63. Limited motion	73. True	83. Abducts	93. True
54. Hypo-thermia	64. Shoulder girdle	74. True	84. True	94. True
55. Supralateral	65. True	75. Deltoid tuber.	85. True	95. True
56. True	66. True	76. True	86. True	96. True
57. True	67. True	77. True	87. True	97. Under
58. True	68. True	78. True	88. True	98. True
59. True	69. True	79. True	89. True	99. True
60. Bacitracin	70. True	80. Hinge	90. True	100. True

CHAPTER 2

LAMINECTOMY CASE STUDY

Laminectomy

This morning the first case on the board for us is a laminectomy for a herniated disk. I will need to stop by x-ray and pick up the patient's folder which contains x-rays, cat scans, MRIs and myelograms. On my way back I will check the chart for any known allergies, previous surgeries, or any other abnormalities. I will also speak to the patient, identify myself and ask if there are any questions that he might have regarding the surgery. I will also check for things that the surgical team should be aware. When I get to the room, I will then check that the room is ready for the patient: x-rays hung, proper drugs available, the case is open and all instrumentation is on hand and working properly, anesthesia has spoken to the patient and we are ready to bring the patient to the room.

While the anesthesiologist is sedating the patient I will stand by and assist with intubations by applying cricoid pressure if needed. With the patient asleep I can now place the Foley, tape the eyes, and prepare the pads for the elbows as we prepare to flip the patient into prone position. With the help of all the team members we will slowly move the patient to the edge, and roll the patient over on to the Wilson frame while anesthesia protects the head and airway. Once over I will place the arms on padded arm boards above the head. The head will be placed in a headrest and padding will then be placed under the knees and feet. Once the patient is fixed properly in the Wilson frame I will check the genitals and arrange the Foley to hang at the head of the table.

Now I will need to place the bovie pad over a fatty location but still close to incision site. I will carefully place warm blankets over the arms and legs of my patient to prevent any hypothermia. The table strap is placed last to prevent the patient from a fall when the table is repositioned. Now I will place a hypo in to the site and get a preoperative x-ray to verify that we are over the proper disc space. With the site properly marked I can beep the surgeon and go scrub while the circulator does the prep. A surgical timeout is done as to recognize the proper patient with their surgery and we can begin.

We will make our incision midline above the affected disk and place a small Beckman self retaining retractor. With the scalpel and the bovie we will dissect down to the paraspinous muscles which we will then take off of the spinous processes and lamina with our Cobb elevator. While the surgeon is doing this, I will make up some hip sponges out of ray-tecs to use as blunt dissectors to free up muscle from bone to gain better exposure.

Now that we have maximum exposure we place a William's retractor in the wound. The spinous processes are cut off flush with the vertebra with the bone scissors, ronjeurs, and Stryker TPS system; all the while I am keeping the instruments clean from bits of bone. Now I will use a small Love root retractor to retract the dura and nerve root while the surgeon cuts away the extruded herniated disk with the help of pituitary ronjeurs and a small up biting curette, being careful not to damage any nerve roots or the aorta, which lies directly under the vertebral column along with the vena cava. While we are working we will use various hemostasis agents like bone wax, gelfoam, surgicele, or thrombin on patties. We will be sure that the patties are removed at the end of the case when the sponge count is initialized. Last we are going to irrigate with bacitracin in our warm saline and prepare to close.

The surgeon will probably leave now and leave me with placement of the drains and closure of the wound. For my closure I will use interrupted 0-vicryl pop offs for the first two layers, then switch to 3-0 vicryl pop offs for the final dermal layer, staples for the skin. Now we are ready for the dressing which will consist of four by fours and a tegaderm. Last we will connect the drain tubes up and tape those down to the patient. Now we should be ready to carefully move the patient back on to his stretcher for extubation. Depending on the reason for the laminectomy the out come should be favorable with the patient able to return to full motion in four to six weeks.

Case study

In this section you will need to describe a laminectomy in detail from your perspective (one that you have done in the past). List any differences in positioning, drugs used, incision, etc.

1. Which tests results should always be in the room at the start of any case?

    A. Lab results

    B. X-rays, cat scans

    C. MRIs, myelograms

    D. All the above

2. The name of the frame for positioning of laminectomy patients is

    A. Packs

    B. Demi

    C. Baker

    D. Wilson

3. The name of the self-retaining retractor used for a laminectomy

    A. Jackson

    B. Demi

    C. Baker

    D. Beckman

4. Which muscle is dissected off the spinous processes during a lamin-ectomy?

    A. Paraspinous

    B. Levator

    C. Splenius

    D. Gluteus maximus

5. A neurosurgical elevator is

   A. Jackson

   B. Key

   C. Cobb

   D. Woodson

6. A neurosurgical nerve root retractor

   A. Love

   B. Key

   C. Penfield

   D. Woodson

7. Large vesicles that lie directly against the vertebral column are

   A. Uterine arteries

   B. Iliolumbars

   C. Aorta, vena cava

   D. Common iliacs

8. A neurosurgical hemostasis agent used with gelfoam is

   A. Thrombin

   B. Bacitracin

   C. Surgicel

   D. Heparin

9. A neurosurgical irrigation agent used on bone is

    A. Gentamicin

    B. Bacitracin

    C. Vancomycin

    D. Ancef

10. The reason for doing a lamin-ectomy is

    A. Habitual disk

    B. Muscle cramps

    C. Back pain

    D. Herniated disk

11. Spondylosis, Stenosis, and Spondyolisthesis most often occur at

    A. T12 and L1

    B. L4-L5 and L5-S1

    C. L1-L4

    D. None of the above

12. Pain from herniation of a lumbar disk may radiate to

    A. The upper arm

    B. Buttock

    C. Posterior thigh

    D. B and C

13. A primary neoplasm may arise from the

A. Meninges

B. Herniated disks

C. Cord compressions

D. Spinal narrowing

14. Pain from herniation of a lumbar disk may radiate to other parts of the body and is called

A. Abnormal

B. Herniated disk

C. Radiculopathy

D. Osteomylitsis

15. This test is used to evaluate spinal injuries

A. X-ray

B. Myleography

C. CT and MRI

D. CT

16. The bony structure of the vertebral column extends from the foramen magnum to

A. Coccyx

B. Sacrum

C. Lower lumbar

D. Filum terminale

17. There are______ number of vertebra in spinal column

A. 28

B. 34

C. 1, 8, and 5

D. 7, 12, and 5

18. The intervertebral disk is held between each vertebra by an annulus fibrous and a(n)

A. Meninges

B. Vertebral arch

C. Posterior ligament

D. Anterior ligament

19. All vertebrae have ________ posterior lateral pedicles

A. Three

B. Two

C. Four

D. Each section different

20. The sacrum has _____ fused sacral vertebrae

A. Two

B. Nine

C. Six

D. Five

21. The coccyx has ______ fused coccygeal vertebrae

    A. Six

    B. Two

    C. Four

    D. Nine

22. The spinal cord is between ______ inches long

    A. 17 and 20

    B. 20 and 25

    C. 12 and 18

    D. 24 and 30

23. There are ______ pairs of spinal nerve roots that branch off each side of the vertebral column

    A. Sixty

    B. Thirty-seven

    C. Eighteen

    D. Thirty-one

24. Vertebrae have a circular canal to house the

    A. Spinal cord

    B. Spinal vessels

    C. Spinal nerves

    D. Spinal ligaments

25. In the cervical region, C1 is the
    A. Coccyx
    B. Sacrum
    C. Axis
    D. Atlas

26. In the cervical region, C2 is the
    A. Coccyx
    B. Sacrum
    C. Axis
    D. Atlas

27. The largest vertebrae lie in the
    A. Lumbar
    B. Thoracic
    C. Cervical
    D. B and C

28. Which vertebral group supports the major weight of the body?
    A. Lumbar
    B. Thoracic
    C. Cervical
    D. B and C

29. Which part of the vertebral column is considered to be part of the pelvic girdle?

A. Coccyx

B. Sacrum

C. Axis

D. Atlas

30. These four fused vertebrae serve as an attachment for several hip and pelvic muscles

A. Coccyx

B. Sacrum

C. Axis

D. Atlas

31. The lateral vertebral extensions are called the

A. Transverse processes

B. Vertebral arch

C. Posterior ligament

D. Anterior ligament

32. The articulating surfaces on a vertebral body are called

A. Meninges

B. Vertebral arch

C. Posterior ligament

D. Facets

33. Cottonoid patties should only be moistened with

A. Spray Thrombin

B. Saline

C. Lactated Ringers

D. All of the above

34. We apply Microfibrillar collagen to the wound site

A. Wet

B. Dry

C. Moist

D. A or B

35. _____ innervates the anterolateral abdominal wall, external genitals, and portions of the lower extremities

A. C5-C8-T1

B. L1-L4

C. L4–L5

D. S1-S4

36. _____ innervates the shoulder and upper limbs

A. C5-C8-T1

B. L1-L4

C. L4–L5

D. S1-S4

37._____ innervates the buttocks, perineum, and lower extremities

A. C5-C8-T1

B. L1-L4

C. L4–L5

D. S1-S4

38. Which surgical procedure is utilized when there is joint instability or intractable pain caused from the narrowing of the disk space?

A. Lysis of adhesions

B. Fusion

C. Osteotomy

D. Bone aspiration

39. _____ is characterized by an abnormal curvature of the vertebral column, and/or posterior convex in thoracic region in elderly people

A. Lordosis

B. Scoliosis

C. Kyphosis

D. Ossification

40. Paraplegia is more likely to occur with a disk rupture in this vertebral system

A. Sacral

B. Thoracic

C. Lumbar

D. Cervical

41. The most common type of abnormal curvature (convex to the side or lateral) is

    A. Lordosis

    B. Scoliosis

    C. Kyphosis

    D. Ossification

42. Swayback or saddleback commonly occurs with pregnancy and is termed

    A. Lordosis

    B. Scoliosis

    C. Kyphosis

    D. Spina Bifida

43. The most common spinal malformation, which occurs in at least 10% of people and is marked by a tuff of hair

    A. Lordosis

    B. Scoliosis

    C. Kyphosis

    D. Spina Bifida

44. The muscle that connects the upper limbs to the trunk is

    A. Trapezius

    B. Latissimus dorsi

    C. Serratus posterior

    D. A and B

45. The muscle of the back that is considered superficial to the respiratory system

A. Trapezius

B. Latissimus dorsi

C. Serratus posterior

D. A and B

46. Arterial blood supply vessels for the spinal cord are reinforced along the way by this tissue

A. Accessory

B. Segmental

C. Radicular

D. Sulcal

47. The three anterior and three posterior veins that drain the spinal cord are referred to as _____ veins

A. Accessory

B. Segmental

C. Spinal

D. Sulcal

48. The innermost covering membrane of the spinal cord is

A. Pia

B. Dura

C. Arachnoid

D. Subdural

49. The outermost covering of the spinal cord is

A. Pia

B. Dura

C. Arachnoid

D. Subdural

50. Fluid received from a lumbar puncture (spinal tap) between these two lumbar vertebra

A. L1-L2

B. L2-L3

C. L3-L4

D. L4-L5

51. All lab tests should be in the room while a surgery is being performed

A. True

B. False

52. A positioning frame for prone patients is a Wilson frame

A. True

C. False

53. A self-retaining retractor used during a laminectomy is called a Baker retractor

A. True

B. False

54. The Paraspinous muscle is dissected off of the spinous process during a laminectomy

    A. True

    B. False

55. A Cobb elevator is mostly used to remove periosteum from bone

    A. True

    B. False

56. A Penfield is a nerve root retractor

    A. True

    B. False

57. The large vesicles that lie directly on the vertebral column are the common iliacs

    A. True

    B. False

58. Heparin is well known hemostasis agent

    A. True

    B. False

59. Bacitracin is well known bone antibiotic

    A. True

    B. False

60. A herniated disk is the main reason for performing a laminectomy

A. True

B. False

61. Spondylosis, Stenosis, and Spondyolisthesis most often occur at L4-L5 and L5-S1

A. True

B. False

62. Pain from herniation of a lumbar disk may radiate to the buttock and the posterior thigh

A. True

B. False

63. Neoplasm's arise off the meninges

A. True

B. False

64. Radiculopathy is where pain from herniation of a lumbar disk radiates to other parts of the body

A. True

B. False

65. An x-ray may be used to evaluate spinal injuries

A. True

B. False

66. The bony structure of the vertebral column extends from the foramen magnum to Coccyx

A. True

B. False

67. There are 34 vertebras in the spinal column

A. True

B. False

68. The intervertebral disk is held between each vertebra by an annulus fibrosis and posterior ligament

A. True

B. False

69. All vertebrae have four posterior lateral pedicles

A. True

B. False

70. The sacrum has five fused sacral vertebrae

A. True

B. False

71. The coccyx has four fused coccygeal vertebrae

A. True

B. False

72. The spinal cord is between 17 and 20 inches long

A. True

B. False

73. There are thirty-one pairs of spinal nerve roots that branch off each side of the vertebral column

A. True

B. False

74. Vertebrae have a circular canal to house the spinal nerves

A. True

B. False

75. In the cervical region, C1 is the atlas

A. True

B. False

76. In the cervical region, C2 is the axis

A. True

B. False

77. The largest vertebrae lie in the lumbar region

A. True

B. False

78. The vertebral group that supports the major weight of the body is the lumbar

A. True

B. False

79. The sacrum is considered to be part of the vertebral column

A. True

B. False

80. The Coccyx is four fused vertebrae which serve as an attachment for several hip and pelvic muscles

A. True

B. False

81. The lateral vertebral extensions are called vertebral arches

A. True

B. False

82. The articulating surfaces on a vertebral body are called facets

A. True

B. False

83. Cottonoid patties should only be moistened with Heparin

A. True

B. False

84. For proper hemostasis, we always apply Microfibrillar collagen to the wound site dry

A. True

B. False

85. L1-L4 innervates the anterolateral abdominal wall, external genitals, and portions of the lower extremities

A. True

B. False

86. C5-C8-T1 innervates the shoulder and upper limbs

A. True

B. False

87. S1-S4 innervates the buttocks, perineum, and lower extremities

A. True

B. False

88. Scoliosis is the surgical procedure that is utilized when there is joint instability or intractable pain caused from the narrowing of the disk space

A. True

B. False

89. Scoliosis is characterized by an abnormal curvature of the vertebral column, and/or posterior convex in thoracic region in elderly people

A. True

B. False

90. Paraplegia is more likely to occur with a disk rupture in the cervical vertebral system

A. True

B. False

91. Most common type of abnormal curvature (convex to the side or lateral) is Scoliosis

A. True

B. False

92. Swayback or saddleback commonly occurs with pregnancy and is termed Lordosis

A. True

B. False

93. The most common spinal malformation, which occurs in at least 10% of people and is marked by a tuff of hair is Spina Bifida

A. True

B. False

94. Muscle that connects the upper limbs to the trunk is Serratus posterior

A. True

B. False

95. Muscle of the back that is considered superficial respiratory is the Latissimus dorsi

A. True

B. False

96. Arterial blood supply vessels for the spinal cord are reinforced along the way by radicular tissue

A. True

B. False

97. The three anterior and three posterior veins that drain the spinal cord are referred to as spinal veins

A. True

B. False

98. The innermost covering membrane of the spinal cord is the dura

A. True

B. False

99. The outermost covering membrane of the spinal cord is the dura

A. True

B. False

100. Fluid is received from a lumbar puncture (spinal tap) between vertebras L3-L4

A. True

B. False

1 A ❑ B ❑ C ❑ D ❑
2 A ❑ B ❑ C ❑ D ❑
3 A ❑ B ❑ C ❑ D ❑
4 A ❑ B ❑ C ❑ D ❑
5 A ❑ B ❑ C ❑ D ❑
6 A ❑ B ❑ C ❑ D ❑
7 A ❑ B ❑ C ❑ D ❑
8 A ❑ B ❑ C ❑ D ❑
9 A ❑ B ❑ C ❑ D ❑
10 A ❑ B ❑ C ❑ D ❑
11 A ❑ B ❑ C ❑ D ❑
12 A ❑ B ❑ C ❑ D ❑
13 A ❑ B ❑ C ❑ D ❑
14 A ❑ B ❑ C ❑ D ❑
15 A ❑ B ❑ C ❑ D ❑
16 A ❑ B ❑ C ❑ D ❑
17 A ❑ B ❑ C ❑ D ❑
18 A ❑ B ❑ C ❑ D ❑
19 A ❑ B ❑ C ❑ D ❑
20 A ❑ B ❑ C ❑ D ❑
21 A ❑ B ❑ C ❑ D ❑
22 A ❑ B ❑ C ❑ D ❑
23 A ❑ B ❑ C ❑ D ❑
24 A ❑ B ❑ C ❑ D ❑
25 A ❑ B ❑ C ❑ D ❑
26 A ❑ B ❑ C ❑ D ❑
27 A ❑ B ❑ C ❑ D ❑
28 A ❑ B ❑ C ❑ D ❑
29 A ❑ B ❑ C ❑ D ❑
30 A ❑ B ❑ C ❑ D ❑
31 A ❑ B ❑ C ❑ D ❑
32 A ❑ B ❑ C ❑ D ❑
33 A ❑ B ❑ C ❑ D ❑
34 A ❑ B ❑ C ❑ D ❑
35 A ❑ B ❑ C ❑ D ❑
36 A ❑ B ❑ C ❑ D ❑
37 A ❑ B ❑ C ❑ D ❑
38 A ❑ B ❑ C ❑ D ❑
39 A ❑ B ❑ C ❑ D ❑
40 A ❑ B ❑ C ❑ D ❑
41 A ❑ B ❑ C ❑ D ❑
42 A ❑ B ❑ C ❑ D ❑
43 A ❑ B ❑ C ❑ D ❑
44 A ❑ B ❑ C ❑ D ❑
45 A ❑ B ❑ C ❑ D ❑
46 A ❑ B ❑ C ❑ D ❑
47 A ❑ B ❑ C ❑ D ❑
48 A ❑ B ❑ C ❑ D ❑
49 A ❑ B ❑ C ❑ D ❑
50 A ❑ B ❑ C ❑ D ❑
51 A ❑ B ❑ C ❑ D ❑
52 A ❑ B ❑ C ❑ D ❑
53 A ❑ B ❑ C ❑ D ❑
54 A ❑ B ❑ C ❑ D ❑
55 A ❑ B ❑ C ❑ D ❑
56 A ❑ B ❑ C ❑ D ❑
57 A ❑ B ❑ C ❑ D ❑
58 A ❑ B ❑ C ❑ D ❑
59 A ❑ B ❑ C ❑ D ❑
60 A ❑ B ❑ C ❑ D ❑
61 A ❑ B ❑ C ❑ D ❑
62 A ❑ B ❑ C ❑ D ❑
63 A ❑ B ❑ C ❑ D ❑
64 A ❑ B ❑ C ❑ D ❑
65 A ❑ B ❑ C ❑ D ❑
66 A ❑ B ❑ C ❑ D ❑
67 A ❑ B ❑ C ❑ D ❑
68 A ❑ B ❑ C ❑ D ❑
69 A ❑ B ❑ C ❑ D ❑
70 A ❑ B ❑ C ❑ D ❑
71 A ❑ B ❑ C ❑ D ❑
72 A ❑ B ❑ C ❑ D ❑
73 A ❑ B ❑ C ❑ D ❑
74 A ❑ B ❑ C ❑ D ❑
75 A ❑ B ❑ C ❑ D ❑
76 A ❑ B ❑ C ❑ D ❑
77 A ❑ B ❑ C ❑ D ❑
78 A ❑ B ❑ C ❑ D ❑
79 A ❑ B ❑ C ❑ D ❑
80 A ❑ B ❑ C ❑ D ❑
81 A ❑ B ❑ C ❑ D ❑
82 A ❑ B ❑ C ❑ D ❑
83 A ❑ B ❑ C ❑ D ❑
84 A ❑ B ❑ C ❑ D ❑
85 A ❑ B ❑ C ❑ D ❑
86 A ❑ B ❑ C ❑ D ❑
87 A ❑ B ❑ C ❑ D ❑
88 A ❑ B ❑ C ❑ D ❑
89 A ❑ B ❑ C ❑ D ❑
90 A ❑ B ❑ C ❑ D ❑
91 A ❑ B ❑ C ❑ D ❑
92 A ❑ B ❑ C ❑ D ❑
93 A ❑ B ❑ C ❑ D ❑
94 A ❑ B ❑ C ❑ D ❑
95 A ❑ B ❑ C ❑ D ❑
96 A ❑ B ❑ C ❑ D ❑
97 A ❑ B ❑ C ❑ D ❑
98 A ❑ B ❑ C ❑ D ❑
99 A ❑ B ❑ C ❑ D ❑
100 A ❑ B ❑ C ❑ D ❑

CH2 Laminectomy

Multiple Choice

1 .D	11. B	21. C	31. A	41. B
2. D	12. D	22. A	32. D	42. A
3. D	13. A	23. D	33. D	43. D
4. A	14. C	24. A	34. B	44. D
5. C	15. C	25. D	35. B	45. B
6. A	16. A	26. C	36. A	46. C
7. C	17. D	27. A	37. D	47. C
8. A	18. C	28. A	38. B	48. A
9. B	19. B	29. B	39. C	49. B
10. D	20. D	30. A	40. B	50. C

True / False

51. True	61. True	71. True	81. Transverse Processes	91. True
52. True	62. True	72. True	82. True	92. True
53. Beckman	63. True	73. True	83. Facets	93. True
54. True	64. True	74. Spinal cord	84. True	94. Trapezius, Serratus
55. True	65. CT, MRI	75. Atlas	85. True	95. True
56. Love	66. True	76. Axis	86. True	96. True
57. Aorta, Vena, cava	67. 7, 12, 5	77. True	87. True	97. True
58. Thrombin	68. True	78. True	88. Fusion	98. Pia
59. True	69. Two	79. True	89. True	99. True
60. True	70. True	80. True	90. Thoracic	100. True

CHAPTER 3

CRANIOTOMY CASE STUDY

Craniotomy

This morning at 3:00am I was called in for an emergency craniotomy, for a subdural hematoma. I will need to stop by x-ray and pick up the patients folder containing x-rays, cat scans, and MRIs. On my way back I will check the chart for any known allergies, previous surgeries, or any other abnormalities. I will also speak to the patient's family, identify myself and ask if there are any questions that he might have regarding the surgery. I will make the team aware of anything extra that they may need. Once I get to the room, I will then check that the room is ready for the patient: x-rays hung, proper drugs available, the case is open and all instrumentation is on hand and working properly, anesthesia has spoken to the patient and we are ready to bring the patient to the room.

While the anesthesiologist is sedating the patient I will stand by and assist with intubations by applying cricoid pressure if needed. With the patient asleep I can now place the Foley, tape the eyes, shave the patient's head. The Mayfield "horseshoe" headrest is applied with Neosporin on the pins, and then the pins are carefully tightened down on to the skull. I will hold the headrest and with the help of all the team members we will slowly move the patient to the edge, and roll the patient over as we are turning the patient laterally then with the patient in proper position, I will lock the headrest down securely. I will drape out the head with 10-10 drapes before prepping. All the while anesthesia protects the head and airway. Once over, I will now place the arms on padded arm boards with pillows between the arms and knees. Once the patient is fixed properly, I will check genitals and arrange the Foley to hang at the foot of the table.

Now I will need to place the bovie pad over a fatty location but still close to incision site. I will carefully place warm blankets over the arms and legs of my patient to prevent any hypothermia. The table strap is placed last to prevent the patient from a fall when the table is repositioned. I will check that the microscope is now setup for the surgeon and me. With the site properly marked I can beep the surgeon and go scrub while the circulator does the prep. After all draping and the throwing off of all cords to be hooked up, we can proceed with the case. A surgical timeout is done as to recognize the proper patient with their surgery and we can begin.

We will make our incision and place a small Beckman self-retaining retractor, with the scalpel and the bovie we will dissect down to the skull, applying Raney clips as we go for hemostasis. When we get to the periosteum we will use our bovie, scalpel, and Cobb elevator. While the surgeon is doing this, I will start getting thrombin soaked patties ready to apply to the scalp edges. We can use hooks now to retract for us while we start to drill.

Now that we have maximum exposure we can begin to drill with a craniotome and Stryker TPS system, all the while I am keeping the instruments clean from bits of bone with antibiotic solution. I will use a number three Penfield to carefully dissect the dura from the skull. Once the bone is removed, we can incise the dura. Currently the surgeon will use a 4-0 silk suture to retract the dura. While the surgeon places the suture, I will hang hemostats on each to retract the durra. Being careful not to damage any nerves or vessels while we are working, we will use various hemostasis agents like bone wax, gelfoam, surgicel, or thrombin on patties. We will be sure that the patties are removed at the end of the case when the sponge count is initialized. We will now apply the Layla as a brain retractor. Now that we have exposure with our Layla, we see that a clot has formed. With a number four Penfield the clot can easily be dissected away. Now the bleeder can easily be taken care of with the bipolar. Last we are going to irrigate with bacitracin in our warm saline and prepare to close.

The closing of dura is first and is done with 4-0 silk pop-offs. Next we will drill small holes in the surrounding skull, and in the piece that we extracted so that a number one silk can be passed through the holes and secured. Drains can be placed now to allow for any oozing of blood in our closure.

The surgeon will probably leave now and leave me with placement of the drains and closure of the wound. For my closure I will use interrupted 0-vicryl pop offs for the first layer, then switch to 3-0 vicryl pop offs for the final dermal layer, and finally staples for the skin. Now we are ready for the dressing which will consist of four by fours and a tegaderm. We will connect the drain tubes and tape those down to the patient. Now we should be ready to carefully move the patient back on to his stretcher for extubation. Depending on the reason for the craniotomy the outcome should be favorable with the patient able to return to full capabilities in four to six weeks.

Cranial nerves

Number	*Name*	*Action*
I.	Olfactory	Smell
II.	Optic	Vision
III.	Oculomotor	Motor and sensory Eyeball, iris, cilary body Proprioception
IV.	Trochlear	Motor and sensory Eyeballs, Proprioception
V.	Trigeminal	Motor and sensory Movement of eyeball Proprioception
VI.	Abducens	Somatic motor Movement of eyeball Proprioception
VII.	Facial	Motor and sensory Facial expression, tear, Saliva secretion
VIII.	Vestibulocochlear	Hearing and balance
IX.	Glossopharyngeal	Visceral motor and sensory Blood pressure regulation, Taste, saliva secretion Proprioception
X.	Vagus	Visceral motor and sensory Sensation from visceral Organs, smooth muscle Contraction and regulation Secretion of digestive fluid
XI.	Spinal accessory	Motor and sensory Proprioception, swallowing, Head movements
XII.	Hypoglossal	Motor and sensory Proprioception Tongue movement, Swallowing

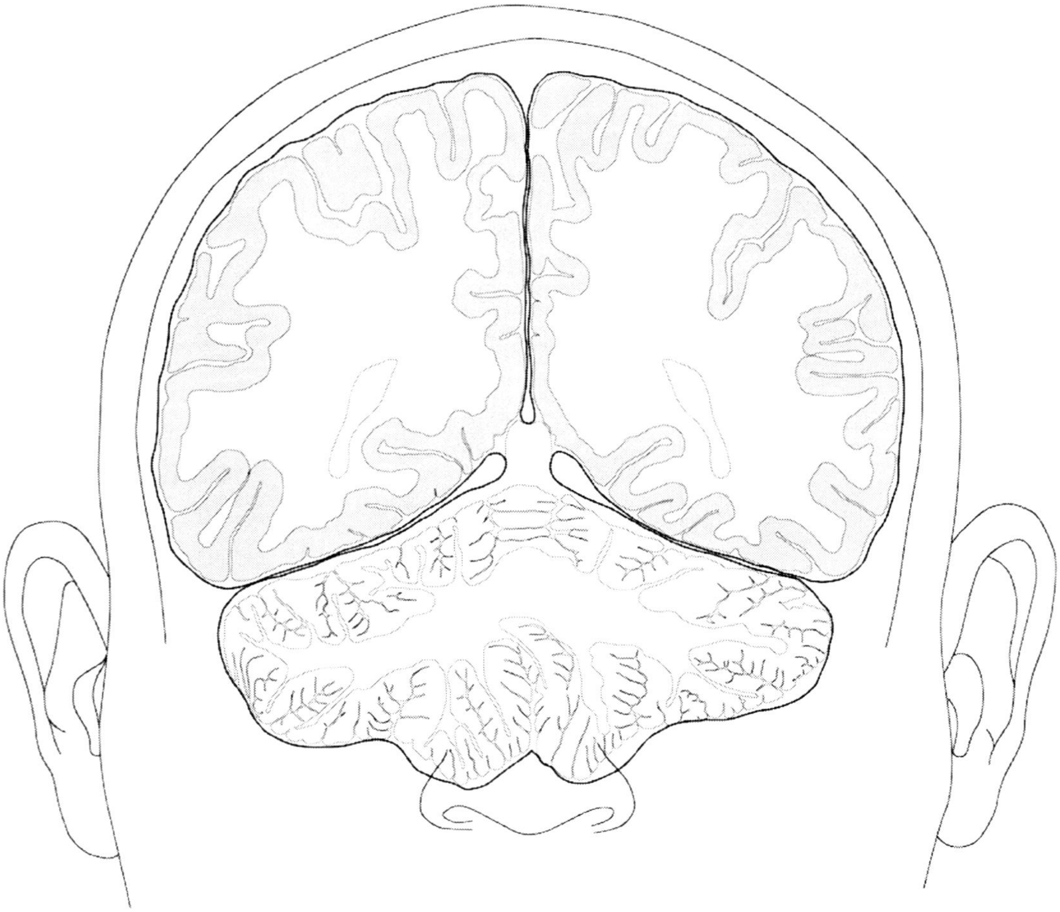

3.1 Anterior brain

Case study

In this section you will need to describe a craniotomy in detail from your perspective (one that you have done in the past). List any differences in positioning, drugs used, or incision, etc.

1. A horseshoe headrest is one name for the ______retractor
    A. Head
    B. Mayfield
    C. Center
    D. Head pad

2. We do need to use a bovie grounding pad with this instrument
    A. Ultra sound
    B. Bayonet forceps
    C. Bipolar forceps
    D. Bovie pencil

3. A disposable gun for scalp hemostasis is called
    A. Hemogun
    B. Raney clip
    C. Clip applier
    D. B and C

4. The craniotome uses this attachment to drill bone
    A. Drill guard
    B. Burr
    C. Perforator
    D. Skull pin

5. These pins are applied to a headrest and will perforate the skin
    A. Drill guard
    B. Burr
    C. Perforator
    D. Skull pin

6. Penfield instruments are used in neuro surgery and are
    A. Elevators
    B. Dissectors
    C. Forceps
    D. A and B

7. A common retractor for brain tissue is
    A. Weitlaner
    B. Shadow
    C. Layla
    D. Beckman

8. The Vagus nerve will innervate
    A. Smooth muscle
    B. Tongue
    C. Retina
    D. Taste

9. The Accessory nerve innervates
    A. Smooth muscle
    B. Tongue
    C. Head movement
    D. Taste

10. The Hypoglossal nerve innervates
    A. Smooth muscle
    B. Tongue
    C. Head movement
    D. Taste

11. The Vagus nerve is listed as
    A. X
    B. XI
    C. XII
    D. VI

12. The Optic nerve is listed as
    A. I
    B. II
    C. III
    D. IV

13. The Hypoglossal nerve is listed as

    A. X

    B. XI

    C. XII

    D. VI

14. The Facial nerve innervates

    A. Smooth muscle

    B. Tongue

    C. Head movement

    D. Taste

15. The Vestibulocochlear nerve innervates

    A. Hearing

    B. Tongue

    C. Head movement

    D. Taste

16. A brain tumor on the Hypoglossal nerve will cause loss of use of the

    A. Eye

    B. Hearing

    C. Tongue

    D. Vision

17. A brain tumor on the Optic nerve will cause loss of the use of

A Eye

B. Hearing

C. Tongue

D. Vision

18. This percentage accounts for all the deaths in head trauma accidents

A. 50%

B. 25%

C. 70%

D. 30%

19. Lethargy, confusion, and other neural deficits are symptoms of

A. Aneurysm

B. Subdural hematoma

C. Spinal cord injury

D. Disease

20. Uncontrolled hypertension and atherosclerosis lead to

A. Aneurysm

B. Stroke

C. Brain hemorrhage

D. All the above

21. Increased amount of CSF will cause

    A. Aneurysm
    B. Stroke
    C. Brain hemorrhage
    D. Hydrocephalus

22. A device that will emit an ultra sonic energy that emulsifies abnormal tissue and then aspirates the tissue is a(n)

    A. CUSA
    B. Ultra sound
    C. Pulse-a-vac
    D. Laser

23. The bone that forms the forehead and nasal cavities

    A. Vomer
    B. Temporal
    C. Occipital
    D. Frontal

24. The bones that form the lateral sides of the head are

    A. Vomer
    B. Temporal
    C. Occipital
    D. Frontal

25. The bone in the base of the nasal cavity is

A. Vomer
B. Temporal
C. Occipital
D. Frontal

26. The bone at the inferior base of the skull is

A. Vomer
B. Temporal
C. Occipital
D. Frontal

27. The cranium is thinner in these patients

A. Males
B. Females
C. Neither
D. Tumor

28. The cheekbones in the face are called the

A. Zygomatic arch
B. Temporal
C. Occipital
D. Frontal

29. The lines of the skull where bones fuse are

    A. Calvaria

    B. Process

    C. Fusion

    D. Suture

30. The medulla oblongata passes through what opening in the skull

    A. Base opening

    B. Foramen magnum

    C. Calvaria

    D. Occipital

31. Fractures in this area of the skull are most common

    A. Base

    B. Foramen magnum

    C. Calvaria

    D. Occipital opening

32. The cranial vault where the brain is stored is the

    A. Base

    B. Foramen magnum

    C. Calvaria

    D. Occipital cavity

33. The Latin term glabellus refers to

    A. Tubercle
    B. Process
    C. Rounded
    D. Smooth and hairless

34. This fossa contains the hypophysis or pituitary gland

    A. Glabellus
    B. Sella turcica
    C. Calvaria
    D. Occipital cavity

35. The mastoid is part of this bone

    A. Zygomatic arch
    B. Temporal
    C. Occipital
    D. Frontal

36. A horizontal fracture of the Maxillae is termed

    A. Le fort I
    B. Le fort II
    C. Le fort III
    D. Maxillae fracture

37. The nerve that is tested by the anesthesiologist to check for depth of anesthesia is

    A. Lacrimal

    B. Supraorbital

    C. Frontalis

    D. Procerus

38. The nerve that is carefully dissected away during a Parotidectomy is

    A. Lacrimal

    B. Supraorbital

    C. Frontalis

    D. Facial

39. Most common cause of Bell's palsy is damage to this nerve

    A. Lacrimal

    B. Supraorbital

    C. Frontalis

    D. Facial

40. The chief artery of the face (Arises from the carotid) is

    A. Lacrimal

    B. Supraorbital

    C. Frontalis

    D. Facial

41. This artery supplies the scalp with blood

    A. Crista
    B. Facial
    C. Occipital
    D. Carotid

42. This skull bone has a lobe of the brain underneath with the same name

    A. Frontal
    B. Temporal
    C. Occipital
    D. All the above

43. The smallest part of the brain

    A. Diencephalon
    B. Midbrain
    C. Pon
    D. Medulla oblongata

44. This is the central part of the brain, (contains the hypo-thalamus)

    A. Diencephalon
    B. Midbrain
    C. Pon
    D. Medulla oblongata

45. This radiographic study is done to locate tumors of the brain

A. Ventriculography

B. Encephalography

C. Tomography

D. All the above

46. An important anastomosis of four arteries that supply blood to the brain is the

A. Carotid

B. Vertebral

C. Circle of Willis

D. Cervical

47. This disease is characterized by three symptoms tremor, rigid, and bradykinesia

A. Tumor

B. Parkinson

C. Epilepsy

D. Neuralgia

48. An uncontrolled discharge of neurons within the brain would define

A. Tumor

B. Parkinson

C. Epilepsy

D. Neuralgia

49. Healing occurs when tissue is clearly incised and reapproximated and repair occurs without complication in
    A. First intention
    B. Second intention
    C. Delayed closure
    D. Primary closure

50. When a wound is allowed to heal open for 5 days and then it is closed is called
    A. First intention
    B. Second intention
    C. Delayed closure
    D. Primary closure

51. A horseshoe head rest is one name for a Wilson frame
    A. True
    B. False

52. We do need to use a bovie grounding pad with a bipolar instrument
    A. True
    B. False

53. A disposable gun for scalp hemostasis is called a Raney clip applier
    A. True
    B. False

54. The craniotome uses a perforator attachment to drill bone

A. True

B. False

55. Perforator pins are applied to a headrest and will perforate the skin

A. True

B. False

56. Penfield instruments are used in neuro surgery and are dissectors and elevators

A. True

B. False

57. A common retractor for brain tissue is the Weitlaner

A. True

B. False

58. The Vagus nerve will innervate smooth tissue

A. True

B. False

59. The Accessory nerve innervates head movement

A. True

B. False

60. The Hypoglossal nerve innervates the tongue

A. True

B. False

61.The Vagus nerve is listed as number V

A. True

B. False

62. The Optic nerve is listed as Number V

A. True

B. False

63. The Hypoglossal nerve is listed as number XII

A. True

B. False

64. The Facial nerve innervates Taste

A. True

B. False

65. The Vestibulocochlear nerve innervates hearing

A. True

B. False

66. A brain tumor on the Hypoglossal nerve will cause loss of use of vision

A. True

B. False

67. A brain tumor on the Optic nerve will cause loss of the use of an eye

A. True

B. False

68. 50 percent of all head trauma accidents will result in death

    A. True

    B. False

69. Lethargy, confusion, and other neural deficits are symptoms of subdural hematoma

    A. True

    B. False

70. Uncontrolled hypertension and atherosclerosis lead to aneurysm, stroke, or brain hemorrhage

    A. True

    B. False

71. Increased amount of CSF will cause hydrocephalus

    A. True

    B. False

72. The CUSA is a device that will emit an ultrasonic energy that emulsifies abnormal tissue and then aspirates the tissue

    A. True

    B. False

73. The bone that forms the forehead and nasal cavities is the Vomer

    A. True

    B. False

74. The bones that form the lateral sides of the head are the temples

    A. True

    B. False

75. The bone in the base of the nasal cavity is the Vomer

    A. True

    B. False

76. The bone at the inferior base of the skull is Occipital bone

    A. True

    B. False

77. The cranium is thinner in female patients

    A. True

    B. False

78. The cheekbones in the face are called the Zygomatic arches

    A. True

    B. False

79. The lines of the skull where bones fuse are called fusion lines

    A. True

    B. False

80. The medulla oblongata passes through the Foramen magnum an opening in the skull

A. True

B. False

81. Fractures in the base of the skull are most common

A. True

B. False

82. The cranial vault where the brain is stored is called the Calvaria

A. True

B. False

83. The Latin term Glabellus refers to smooth and hairless

A. True

B. False

84. The Sella Turcica fossa contains the hypophysis or pituitary gland

C. True

D. False

85. The mastoid is part of the Occipital bone

A. True

B. False

86. A horizontal fracture of the Maxillae is termed a Le fort II

    A. True

    B. False

87. The nerve that is tested by the anesthesiologist to check for depth of anesthesia is the Supraorbital

    A. True

    B. False

88. The nerve that is carefully dissected away during a Parotidectomy is the Facial nerve

    A. True

    B. False

89. The most common cause of Bell's palsy is damage to the Facial nerve

    A. True

    B. False

90. The chief artery of the face, the Facial, arises from the carotid

    A. True

    B. False

91. The carotid artery supplies the scalp with blood

    A. True

    B. False

92. Most skull bones have a lobe of the brain underneath with the same name as the bone

A. True

B. False

93. The Pons is the smallest part of the brain

A. True

B. False

94. The Pons is the central part of the brain, (contains the hypo-thalamus)

A. True

B. False

95. An Encephalography radiographic study is just one of the tests done to locate tumors of the brain

A. True

B. False

96. The Circle of Willis is an important anastomosis of four arteries that supply blood to the brain

A. True

B. False

97. Epilepsy disease is characterized by three symptoms tremor, rigid, and bradykinesia

A. True

B. False

98. Epilepsy is uncontrolled discharge of neurons within the brain

   A. True

   B. False

99. Healing occurs when tissue is clearly incised and reapproximated and repair occurs without complication in first intention

   A. True

   B. False

100. When a wound is allowed to heal open for 5 days and then is closed is called second intention

   A. True

   B. False

1 A ❑ B ❑ C ❑ D ❑
2 A ❑ B ❑ C ❑ D ❑
3 A ❑ B ❑ C ❑ D ❑
4 A ❑ B ❑ C ❑ D ❑
5 A ❑ B ❑ C ❑ D ❑
6 A ❑ B ❑ C ❑ D ❑
7 A ❑ B ❑ C ❑ D ❑
8 A ❑ B ❑ C ❑ D ❑
9 A ❑ B ❑ C ❑ D ❑
10 A ❑ B ❑ C ❑ D ❑
11 A ❑ B ❑ C ❑ D ❑
12 A ❑ B ❑ C ❑ D ❑
13 A ❑ B ❑ C ❑ D ❑
14 A ❑ B ❑ C ❑ D ❑
15 A ❑ B ❑ C ❑ D ❑
16 A ❑ B ❑ C ❑ D ❑
17 A ❑ B ❑ C ❑ D ❑
18 A ❑ B ❑ C ❑ D ❑
19 A ❑ B ❑ C ❑ D ❑
20 A ❑ B ❑ C ❑ D ❑
21 A ❑ B ❑ C ❑ D ❑
22 A ❑ B ❑ C ❑ D ❑
23 A ❑ B ❑ C ❑ D ❑
24 A ❑ B ❑ C ❑ D ❑
25 A ❑ B ❑ C ❑ D ❑
26 A ❑ B ❑ C ❑ D ❑
27 A ❑ B ❑ C ❑ D ❑
28 A ❑ B ❑ C ❑ D ❑
29 A ❑ B ❑ C ❑ D ❑
30 A ❑ B ❑ C ❑ D ❑
31 A ❑ B ❑ C ❑ D ❑
32 A ❑ B ❑ C ❑ D ❑
33 A ❑ B ❑ C ❑ D ❑
34 A ❑ B ❑ C ❑ D ❑
35 A ❑ B ❑ C ❑ D ❑
36 A ❑ B ❑ C ❑ D ❑
37 A ❑ B ❑ C ❑ D ❑
38 A ❑ B ❑ C ❑ D ❑
39 A ❑ B ❑ C ❑ D ❑
40 A ❑ B ❑ C ❑ D ❑
41 A ❑ B ❑ C ❑ D ❑
42 A ❑ B ❑ C ❑ D ❑
43 A ❑ B ❑ C ❑ D ❑
44 A ❑ B ❑ C ❑ D ❑
45 A ❑ B ❑ C ❑ D ❑
46 A ❑ B ❑ C ❑ D ❑
47 A ❑ B ❑ C ❑ D ❑
48 A ❑ B ❑ C ❑ D ❑
49 A ❑ B ❑ C ❑ D ❑
50 A ❑ B ❑ C ❑ D ❑
51 A ❑ B ❑ C ❑ D ❑
52 A ❑ B ❑ C ❑ D ❑
53 A ❑ B ❑ C ❑ D ❑
54 A ❑ B ❑ C ❑ D ❑
55 A ❑ B ❑ C ❑ D ❑
56 A ❑ B ❑ C ❑ D ❑
57 A ❑ B ❑ C ❑ D ❑
58 A ❑ B ❑ C ❑ D ❑
59 A ❑ B ❑ C ❑ D ❑
60 A ❑ B ❑ C ❑ D ❑
61 A ❑ B ❑ C ❑ D ❑
62 A ❑ B ❑ C ❑ D ❑
63 A ❑ B ❑ C ❑ D ❑
64 A ❑ B ❑ C ❑ D ❑
65 A ❑ B ❑ C ❑ D ❑
66 A ❑ B ❑ C ❑ D ❑
67 A ❑ B ❑ C ❑ D ❑
68 A ❑ B ❑ C ❑ D ❑
69 A ❑ B ❑ C ❑ D ❑
70 A ❑ B ❑ C ❑ D ❑
71 A ❑ B ❑ C ❑ D ❑
72 A ❑ B ❑ C ❑ D ❑
73 A ❑ B ❑ C ❑ D ❑
74 A ❑ B ❑ C ❑ D ❑
75 A ❑ B ❑ C ❑ D ❑
76 A ❑ B ❑ C ❑ D ❑
77 A ❑ B ❑ C ❑ D ❑
78 A ❑ B ❑ C ❑ D ❑
79 A ❑ B ❑ C ❑ D ❑
80 A ❑ B ❑ C ❑ D ❑
81 A ❑ B ❑ C ❑ D ❑
82 A ❑ B ❑ C ❑ D ❑
83 A ❑ B ❑ C ❑ D ❑
84 A ❑ B ❑ C ❑ D ❑
85 A ❑ B ❑ C ❑ D ❑
86 A ❑ B ❑ C ❑ D ❑
87 A ❑ B ❑ C ❑ D ❑
88 A ❑ B ❑ C ❑ D ❑
89 A ❑ B ❑ C ❑ D ❑
90 A ❑ B ❑ C ❑ D ❑
91 A ❑ B ❑ C ❑ D ❑
92 A ❑ B ❑ C ❑ D ❑
93 A ❑ B ❑ C ❑ D ❑
94 A ❑ B ❑ C ❑ D ❑
95 A ❑ B ❑ C ❑ D ❑
96 A ❑ B ❑ C ❑ D ❑
97 A ❑ B ❑ C ❑ D ❑
98 A ❑ B ❑ C ❑ D ❑
99 A ❑ B ❑ C ❑ D ❑
100 A ❑ B ❑ C ❑ D ❑

CH3 Craniotomy

Multiple Choice

1. B	11. A	21. D	31. A	41. D
2. D	12. A	22. A	32. C	42. D
3. D	13. C	23. D	33. D	43. B
4. C	14. D	24. B	34. B	44. A
5. D	15. A	25. A	35. B	45. D
6. D	16. C	26. C	36. A	46. C
7. C	17. D	27. B	37. B	47. B
8. A	18. A	28. A	38. D	48. C
9. C	19. B	29. D	39. D	49. A
10. B	20. D	30. B	40. D	50. C

True / False

51. Mayfield	61. X	71. True	81. True	91. True
52. True	62. II	72. True	82. True	92. True
53. True	63. True	73. Frontal	83. True	93. Midbrain
54. True	64. True	74. Temporal	84. True	94. Dien-cephalon
55. Skull pins	65. True	75. True	85. Temporal	95. True
56. True	66. Tongue	76. True	86. Le fort I	96. True
57. Layla	67. True	77. True	87. True	97. Parkinson
58. True	68. True	78. True	88. True	98. True
59. True	69. True	79. Suture	89. True	99. True
60. True	70. True	80. True	90. True	100. True

CHAPTER 4

TOTAL HYSTERECTOMY CASE STUDY

Abdominal Hysterectomy

This afternoon the first case on the board for us is an abdominal hysterectomy. I will need to stop by the lab for any pregnancy test results. On my way back I will check the chart for any known allergies, previeious surgeries, or any other abnormalities. I will also speak to the patient, identify myself and ask if there are any questions that she might have regarding the surgery. I will make the team aware of anything extra that they may need. Once I get to the room, I will then check that the room is ready for the patient: Lab tests available, proper drugs available, the case is open and all instrumentation is on hand and working properly, that anesthesia has spoken to the patient and we are ready to bring the patient to the room. The reason we are doing this surgery today is that the patient has abnormal bleeding along with fibroids of the uterus.

While the anesthesiologist is sedating the patient, I will assist with intubations by applying cricoid pressure (Sellecks procedure) if needed. With the patient asleep I can now place the Foley, and since this is a female I will only need to insert the catheter about four inches, making sure that I have urine before inflating the balloon. The patient is placed into supine position. With the help of all the team members we will slowly move our patient to the edge of the table while anesthesia protects the head and airway and move.

Now I will need to place the bovie pad over a fatty location but still close to incision site. Also I should apply T.E.D. hose (antiembolic device) for proper venous circulation. I will carefully place warm blankets over the arms and legs of my patient to prevent any hypothermia. The table strap is placed last to prevent the patient from a fall when the table is repositioned. Since the head of the table will be dropped the patient's knee should be over a break in the bed so the legs can be lowered also. With the site properly marked I can beep the surgeon and go scrub while the circulator does the prep. The Surgical technologist and I will drape the patient, and hook up bovie and suction while the surgeon is scrubbing. A surgical timeout is done as to recognize the proper patient with their surgery and we can begin.

The incision we are going to use is pfannenstiel. We only extend the incision as necessary to get the proper exposure. I will use a small Richardson, as the super fiscal layers of tissue need to be exposed. With use of the bovie the fascia layers are opened transversely on either side of the linea alba next we will grab the fascia layers with two kockers and as I lift up on these instruments the surgeon will dissect the muscle off superiorly and inferiorly. We will then split the muscle fibers with blunt dissection and very carefully open the peritoneum vertically. Once opened the incision can be stretched further with the use of our hands. I will then place two moistened lap sponges at the wound edges under the balfor retractor blades to avert any nerve damage.

Now we can get some Trendelenburg so the bowels will fall away and place three moistened lap sponges in the wound to pack the bowel and gain maximum exposure. Now when the bladder blade is attached I will grab the uterus with a leahe tenaculum as to manipulate it. The first ligaments that we encounter will be the round and ovarian ligaments and will need to be double Kelly clamped, cut and tied off. For our suture we will be using a number 1 Vicryl popoff and a Heaney stitch. Once the surgeon has his side secure, I will proceed to clamp, cut and suture my side. The technique I use is a one handed tie; this permits a secure knot without having to get both hands into the wound.

Now I will make a window through the broad ligament so a curved Heaney clamp can be used. Once the surgeon finishes his side we cut the peritoneum off the anterior and posterior side of the uterus with Metzenbaum scissors and with the help of a sponge stick we will use blunt dissection to remove the bladder flap from the uterus. As the peritoneum is dissected away we must locate the ureters and use constant vigilance as not to damage them during this procedure. Once again we can start double clamping and cutting, and this time double tying as the uterine artery will be in this clamp. Single clamping now can be used as the uterine artery is already tied off.

With having the uterine vessels tied off now the cervix will need to be mobilized, clamped, cut, and tied. In these clamps we will encounter the cardinal ligaments. Now we are ready to double clamp with curved Heanys at the vaginal cuff. We can use Scott Jorgensen scissors for this cut and tie off these vessels. The vaginal cuff will be closed with a number 1 Vicryl running suture. As we exit the surgical site we do one last check for bleeders and ensure that the ureters are left undamaged.

Last, we are going to irrigate with warm saline and prepare to close. For my closure I will use a 2-0 Vicryl runner to close the peritoneum, then a number 1 Vicryl running stitch times two to close the fascia layers, and finishing the closure with a 4-0 Vicryl subque. I will apply steri strips and mastasol; the dressing will be fluffs and medipore tape.

The drapes are then removed and the table will return slowly to level so that the patient can be extubated. Care will be taken as we proceed to move the patient to her stretcher and transported to recovery. On my follow up I will advise the patient on dressing changes, showers, and basic movements. The expected outcome will be to return to normal activities in six weeks.

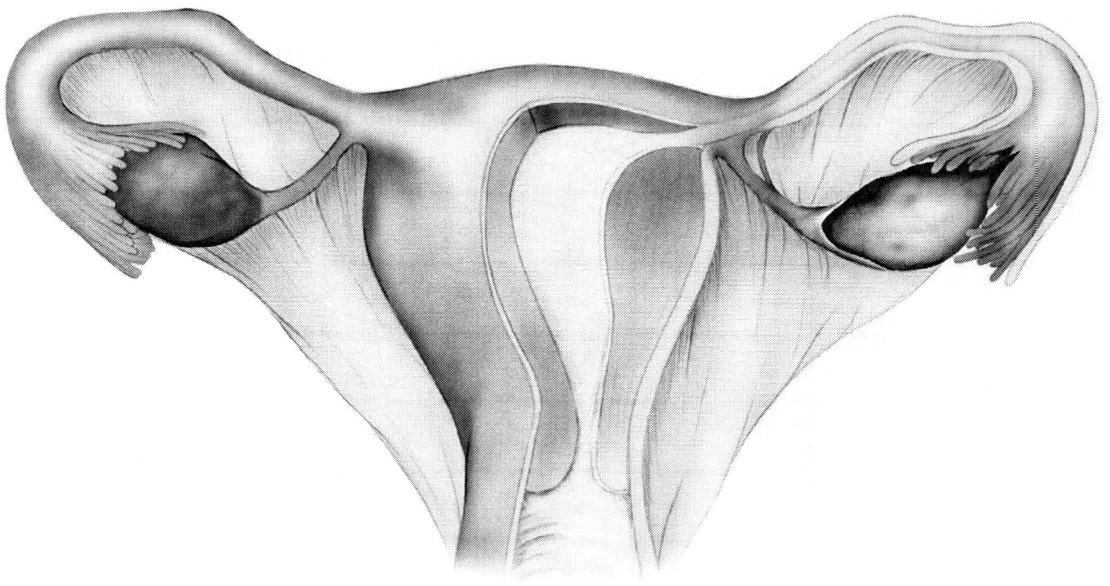

4.1 Uterus

Case study

In this section you will need to describe a hysterectomy in detail from your perspective (one that you have done in the past). List any differences in positioning, drugs used, or incision, etc.

1. Positive results of a pregnancy test should affect a hysterectomy in which way?

   A. Cancel surgery

   B. Postpone surgery

   C. Reschedule surgery

   D. All the above

2. Antiembolic devices (SCD stockings, T.E.D. hose) should be used for this reason

   A. Prevent embolism

   B. Increase blood flow

   C. Proper oxygenation

   D. All of the above

3. One of the most common abdominal hysterectomy incisions is the

   A. Heaney

   B. Maylard

   C. Pfannenstiel

   D. Cherney

4. The least hemorrhagic incision for an abdominal hysterectomy is the

   A. Vertical

   B. Maylard

   C. Pfannenstiel

   D. Cherney

5. The suturing technique for tying off uterine vessels is called

   A. Heaney

   B. Mayo

   C. Maylard

   D. Cherney

6. During a myomectomy procedure, the suturing technique is

   A. Baseball stitch

   B. Football stitch

   C. Heaney stitch

   D. Mayo stitch

7. Instillation of 10% _______ into the surgical wound can reduce adhesions after a myomectomy

   A. Platelets

   B. Ringers

   C. Dextran

   D. Saline

8. Peritoneum sheet that is dissected off the uterus (posterior and anterior leaves) during an abdominal hysterectomy

   A. Round ligament

   B. Broad ligament

   C. Infundibulopelvic ligament

   D. Cardinal ligament

9. Position of patient to allow intestines to fall cephalad is

    A. Lithotomy

    B. Supine

    C. Trendelenburg

    D. Reverse trendelenburg

10. Reason for a vertical incision during an abdominal hysterectomy can be

    A. Fibroids

    B. Cancer

    C. Pervious incision

    D. All of the above

11. The main supporting ligaments of the uterus are the

    A. Round ligaments

    B. Broad ligaments

    C. Infundibulopelvic ligaments

    D. Cardinal ligaments

12. The ilia, ischia, and the pubic bones are considered the

    A. Bony pelvis

    B. Pelvic floor

    C. Pelvic girdle

    D. All of the above

13. The u-shaped muscle that originates at the pubic bone and attaches to lateral walls of the vagina and rectum is

A. Puborectalis

B. Ilia

C. Levator ani

D. Pubococcygeal

14. The sensitive perineum area is innervated by the _______nerve

A. Vesicocervical

B. Pudendal

C. Coccyx

D. Uterine

15. Clamping, cutting, and tying off of vessels and ligaments close to the uterus and cervix is called

A. Marsupialization

B. Skeltonization

C. Paring

D. Heaney

16. Which nerve might be damaged during an episiotomy?

A. Ilioinguinal

B. Dorsal

C. Perineal

D. Pudendal

17. The ovarian arteries arise from which structure

    A. Aorta

    B. Hypogastric

    C. External iliac

    D. Internal iliac

18. The left ovarian vein drains into this structure

    A. Vena cava

    B. Renal vein

    C. External iliac

    D. Internal iliac

19. The term mittelschmerz refers to pain of the

    A. Ovaries

    B. Uterus

    C. Sacrum

    D. Rectum

20. Innervations of the uterus arises from these inferior hypogastric plexus at

    A. L3-L4

    B. L4- L5

    C. S1- S2

    D. S2, S3, and S4

21. The uterine tubes are divided at the ________ during an abdominal hysterectomy

A. Infundibulum

B. Ampulla

C. Isthmus

D. Cornu

22. Remnants of the gubernaculums in embryos are later

A. Ovaries

B. Round ligaments

C. Broad ligaments

D. Uterine tubes

23. The wall of the bladder is composed chiefly of smooth muscle called

A. Membranous

B. Vesical plexes

C. Detrusor

D. Epithelium

24. Innervations of the urinary bladder arise from the pelvicsplanchnic at

A. L3-L4

B. L4- L5

C. S1- S2

D. S2, S3, and S4

25. The trianglular area (smooth muscle) in the urinary bladder is known as

A. Mucus membrane

B. Vesico pouch

C. Trigone

D. Triangle

26. Two structures that lie within the peritoneum sheets known as the broad ligament are

A. Ureters

B. Renal veins

C. External iliac

D. Internal iliac

27. On a pfannenstiel incision, we will pass through how many different layers of tissue

A. Three

B. Six

C. Four

D. Five

28. Most of the pelvis blood is supplied by which main arteries

A. Aorta

B. Internal iliacs

C. External iliacs

D. Uterine artery

29. In the pelvis, the common iliacs will transverse through this ring

    A. Obturator

    B. Femoral

    C. Pectin

    D. Pubic

30. The largest and most important muscles in the pelvic floor are

    A. Coccygeus

    B. Puborectalis

    C. Levator Ani

    D. Obturator internus

31. This fibromuscular diaphragm supports the pelvic floor and viscera

    A. Coccygeus

    B. Puborectalis

    C. Levator Ani

    D. Obturator internus

32. Of the four types of bony pelvises the one best designed for child birth is

    A. Anthropoid

    B. Android

    C. Platypelloid

    D. Gynecoid

33. The medial aspect of the hip bone is formed by the ilium, ischium, and the

____

A. Pubis

B. Coccyx

C. Acetabulum

D. Sacrum

34. To convert a closed cavity (cyst) into an open pouch is called

A. Degustation

B. Deviation

C. Palpate

D. Marsupialization

35. The condition of the bladder protruding into the vagina is known as

A. Rectocele

B. Enterocele

C. Cystocele

D. Proctocele

36. The condition of the intestines protruding in to the vagina is known as

A. Rectocele

B. Enterocele

C. Cystocele

D. Proctocele

37. Irrigation to cause lysis of malignant cancer cells would be

    A. Distilled water

    B. Normal saline

    C. Lactated ringers

    D. DW5

38. An infection that develops while staying in the hospital is considered

    A. Host pathogen

    B. Nosocomial

    C. Spore

    D. Fomite

39. Most incisional infections are caused by

    A. Surgeon

    B. Assistant

    C. Patients flora

    D. Airborne

40. An acute, inflamed, open, fresh wound is considered

    A. Clean

    B. Clean contaminated

    C. Contaminated

    D. Dirty

41. A wound that is left open to heal, because of infection or tissue breakdown is called

A. First intention
B. Second intention
C. Third intention
D. Forth intention

42. A wound that is delayed and closed at a later date is closed by

A. First intention
B. Second intention
C. Third intention
D. Forth intention

43. Which Apgar score would be considered best for a new born infant

A. Two
B. Four
C. Eight
D. A or B

44. Uterine fibroids that arise from the uterine wall usually benign can be eliminated by

A. Marsupialization
B. Myomectomy
C. Vulvectomy
D. Histogram

45. Uterine cancer or fibromas can be eliminated with this procedure

    A. Marsupialization

    B. Myomectomy

    C. Vulvectomy

    D. Hysterectomy

46. Carcinoma in situ of the uterus is curable to what percentage

    A. 75%

    B. 100%

    C. 50%

    D. 90%

47. Excessive or prolonged uterine bleeding is usually considered

    A. Hypermenorrhea

    B. Hypomenorrhea

    C. Polymenorrhea

    D. Intermenstural

48. This is a term that indicates how many times a woman has been pregnant

    A. Braxton Hicks

    B. Station

    C. Parity

    D. Gravida

49. This is a term that indicates how many times a woman has given birth

A. Braxton Hicks

B. Station

C. Parity

D. Gravida

50. An oxytocic drug used to induce or continue labor is

A. Pitocin

B. Hembate

C. Carboprost

D. Digitoxin

51. Positive results of a pregnancy test should cancel surgery

A. True

B. False

52. Antiembolic devices (SCD stockings, T.E.D. hose) should always be used on long surgeries

A. True

B. False

53. Cherney is one of the most common abdominal hysterectomy incisions

A. True

B. False

54. The least hemorrhagic incision for an abdominal hysterectomy is the vertical incision

A. True

B. False

55. The suturing technique for tying off uterine vessels is called a Heaney stitch

A. True

B. False

56. During a myomectomy procedure, the suturing technique is called a baseball stitch

A. True

B. False

57. Instillation of 10% Ringers into the surgical wound can reduce adhesions after a myomectomy

A. True

B. False

58. Peritoneum sheet that is dissected off the uterus (posterior and anterior leaves) during an abdominal hysterectomy is called broad ligament

A. True

B. False

59. Position of patient to allow intestines to fall cephalad is reverse Trendelenburg

A. True

B. False

60. Reason for a vertical incision during an abdominal hysterectomy can be cancer

    A. True

    B. False

61. The main supporting ligaments of the uterus are the Cardinal ligaments

    A. True

    B. False

62. The pelvic girdle is made up of the ilia, ischia, and the pubic bones

    A. True

    B. False

63. The u-shaped muscle that originates at the pubic bone and attaches to lateral walls of the vagina and rectum is the Levator ani

    A. True

    B. False

64. The sensitive perineum area is innervated by the uterine nerve

    A. True

    B. False

65. Clamping, cutting, and tying off of vessels and ligaments close to the uterus and cervix is called skeltonization

    A. true

    B. False

66. The pudendal nerve is at risk of damaged during an episiotomy

    A. True

    B. False

67. The ovarian arteries arise from aorta

    A. True

    B. False

68. The left ovarian vein drains into the vena cava

    A. True

    B. False

69. The term mittelschmerz refers to pain of the ovaries

    A. True

    B. False

70. Innervations of the uterus arises from these inferior hypogastric plexus at S1, S2, and S3

    A. True

    B. False

71. The uterine tubes are divided at the isthmus during an abdominal hysterectomy

    A. True

    B. False

72. Remnants of the gubernaculums in embryos are later round ligaments

A. True

B. False

73. The wall of the bladder is composed chiefly of smooth muscle, called epithelium

A. True

B. False

74. Innervations of the urinary bladder arise from the pelvicsplanchnic at L4-L5

A. True

B. False

75. The trianglular area (smooth muscle) in the urinary bladder is known as the trigone

A. True

B. False

76. Two structures that lie within the peritoneum sheets known as the broad ligament are the ureters

A. True

B. False

77. On a pfannenstiel incision, we will pass through six different layers of tissue

A. True

B. False

78. Most of the pelvis blood is supplied by the internal iliacs

A. True

B. False

79. In the pelvis, the common iliacs will transverse through the femoral ring

A. True

B. False

80. The largest and most important muscles in the pelvic floor are the levator ani

A. True

B. False

81. The fibromuscular diaphragm which supports the pelvic floor and viscera is the levator ani

A. True

B. False

82. Of the four types of bony pelvises the one best designed for child birth is an android

A. True

B. False

83. The medial aspect of the hip bone is formed by the ilium, ischium, and the coccyx

A. True

B. False

84. To convert a closed cavity (cyst) into an open pouch is called marsupialization

A. True

B. False

85. The condition of the bladder protruding in to the vagina is known as enterocele

A. True

B. False

86. The condition of the intestines protruding in to the vagina is known as enterocele

A. True

B. False

87. Irrigation to cause lysis of malignant cancer cells would be with distilled water

A. True

B. False

88. An infection that develops while staying in the hospital is considered nosocomial

A. True

B. False

89. Most incisional infections are caused by the assistant

A. True

B. False

90. An acute, inflamed, open, fresh wound is considered contaminated

A. True

B. False

91. A wound that is left open to heal, because of infection or tissue breakdown is called second intention

A. True

B. False

92. A wound that is delayed and closed at a later date is closed by second intention

A. True

B. False

93. An Apgar score of ten would be considered best for a new born infant

A. True

B. False

94. Uterine fibroids that arise from the uterine wall usually benign can be eliminated by myomectomy

A. True

B. False

95. Uterine cancer or fibromas can be eliminated with a hysterectomy

A. True

B. False

96. Carcinoma in situ of the uterus is curable to 100 percent

A. True

B. False

97. Excessive or prolonged uterine bleeding is usually considered hypermenorrhea

    A. True

    B. False

98. Gravida is a term that indicates how many times a woman has been pregnant

    A. True

    B. False

99. Parity is a term that indicates how many times a woman has given birth

    A. True

    B. False

100. Pitocin is an oxytocic drug used to induce or continue labor

    A. True

    B. False

1 A ☐ B ☐ C ☐ D ☐
2 A ☐ B ☐ C ☐ D ☐
3 A ☐ B ☐ C ☐ D ☐
4 A ☐ B ☐ C ☐ D ☐
5 A ☐ B ☐ C ☐ D ☐
6 A ☐ B ☐ C ☐ D ☐
7 A ☐ B ☐ C ☐ D ☐
8 A ☐ B ☐ C ☐ D ☐
9 A ☐ B ☐ C ☐ D ☐
10 A ☐ B ☐ C ☐ D ☐
11 A ☐ B ☐ C ☐ D ☐
12 A ☐ B ☐ C ☐ D ☐
13 A ☐ B ☐ C ☐ D ☐
14 A ☐ B ☐ C ☐ D ☐
15 A ☐ B ☐ C ☐ D ☐
16 A ☐ B ☐ C ☐ D ☐
17 A ☐ B ☐ C ☐ D ☐
18 A ☐ B ☐ C ☐ D ☐
19 A ☐ B ☐ C ☐ D ☐
20 A ☐ B ☐ C ☐ D ☐
21 A ☐ B ☐ C ☐ D ☐
22 A ☐ B ☐ C ☐ D ☐
23 A ☐ B ☐ C ☐ D ☐
24 A ☐ B ☐ C ☐ D ☐
25 A ☐ B ☐ C ☐ D ☐
26 A ☐ B ☐ C ☐ D ☐
27 A ☐ B ☐ C ☐ D ☐
28 A ☐ B ☐ C ☐ D ☐
29 A ☐ B ☐ C ☐ D ☐
30 A ☐ B ☐ C ☐ D ☐
31 A ☐ B ☐ C ☐ D ☐
32 A ☐ B ☐ C ☐ D ☐
33 A ☐ B ☐ C ☐ D ☐
34 A ☐ B ☐ C ☐ D ☐
35 A ☐ B ☐ C ☐ D ☐
36 A ☐ B ☐ C ☐ D ☐
37 A ☐ B ☐ C ☐ D ☐
38 A ☐ B ☐ C ☐ D ☐
39 A ☐ B ☐ C ☐ D ☐
40 A ☐ B ☐ C ☐ D ☐
41 A ☐ B ☐ C ☐ D ☐
42 A ☐ B ☐ C ☐ D ☐
43 A ☐ B ☐ C ☐ D ☐
44 A ☐ B ☐ C ☐ D ☐
45 A ☐ B ☐ C ☐ D ☐
46 A ☐ B ☐ C ☐ D ☐
47 A ☐ B ☐ C ☐ D ☐
48 A ☐ B ☐ C ☐ D ☐
49 A ☐ B ☐ C ☐ D ☐
50 A ☐ B ☐ C ☐ D ☐
51 A ☐ B ☐ C ☐ D ☐
52 A ☐ B ☐ C ☐ D ☐
53 A ☐ B ☐ C ☐ D ☐
54 A ☐ B ☐ C ☐ D ☐
55 A ☐ B ☐ C ☐ D ☐
56 A ☐ B ☐ C ☐ D ☐
57 A ☐ B ☐ C ☐ D ☐
58 A ☐ B ☐ C ☐ D ☐
59 A ☐ B ☐ C ☐ D ☐
60 A ☐ B ☐ C ☐ D ☐
61 A ☐ B ☐ C ☐ D ☐
62 A ☐ B ☐ C ☐ D ☐
63 A ☐ B ☐ C ☐ D ☐
64 A ☐ B ☐ C ☐ D ☐
65 A ☐ B ☐ C ☐ D ☐
66 A ☐ B ☐ C ☐ D ☐
67 A ☐ B ☐ C ☐ D ☐
68 A ☐ B ☐ C ☐ D ☐
69 A ☐ B ☐ C ☐ D ☐
70 A ☐ B ☐ C ☐ D ☐
71 A ☐ B ☐ C ☐ D ☐
72 A ☐ B ☐ C ☐ D ☐
73 A ☐ B ☐ C ☐ D ☐
74 A ☐ B ☐ C ☐ D ☐
75 A ☐ B ☐ C ☐ D ☐
76 A ☐ B ☐ C ☐ D ☐
77 A ☐ B ☐ C ☐ D ☐
78 A ☐ B ☐ C ☐ D ☐
79 A ☐ B ☐ C ☐ D ☐
80 A ☐ B ☐ C ☐ D ☐
81 A ☐ B ☐ C ☐ D ☐
82 A ☐ B ☐ C ☐ D ☐
83 A ☐ B ☐ C ☐ D ☐
84 A ☐ B ☐ C ☐ D ☐
85 A ☐ B ☐ C ☐ D ☐
86 A ☐ B ☐ C ☐ D ☐
87 A ☐ B ☐ C ☐ D ☐
88 A ☐ B ☐ C ☐ D ☐
89 A ☐ B ☐ C ☐ D ☐
90 A ☐ B ☐ C ☐ D ☐
91 A ☐ B ☐ C ☐ D ☐
92 A ☐ B ☐ C ☐ D ☐
93 A ☐ B ☐ C ☐ D ☐
94 A ☐ B ☐ C ☐ D ☐
95 A ☐ B ☐ C ☐ D ☐
96 A ☐ B ☐ C ☐ D ☐
97 A ☐ B ☐ C ☐ D ☐
98 A ☐ B ☐ C ☐ D ☐
99 A ☐ B ☐ C ☐ D ☐
100 A ☐ B ☐ C ☐ D ☐

CH4 Hysterectomy

Multiple Choice

1. D	11. D	21. C	31. C	41. B
2. D	12. D	22. B	32. D	42. C
3. C	13. A	23. C	33. A	43. C
4. A	14. B	24. D	34. D	44. B
5. A	15. B	25. C	35. C	45. D
6. A	16. D	26. A	36. B	46. B
7. C	17. A	27. B	37. A	47. A
8. B	18. B	28. B	38. B	48. D
9. C	19. A	29. B	39. C	49. C
10. D	20. D	30. C	40. C	50. A

True / False

51. True	61. True	71. True	81. True	91. True
52. True	62. True	72. True	82. Gynecoid	92. Third intention
53. Pfannenstiel	63. True	73. Detrusor	83. Pubis	93. True
54. True	64. Pudendal	74. S2,S3,S4	84. True	94. True
55. True	65. True	75. True	85. Cystocele	95. True
56. True	66. True	76. True	86. True	96. True
57. Dextran	67. True	77. True	87. True	97. True
58. True	68. Renal vein	78. True	88. True	98. True
59. Trendelen-burg	69. True	79. True	89. Patients flora	99. True
60. True	70. True	80. True	90. True	100. True

CHAPTER 5

NEPHRECTOMY CASE STUDY

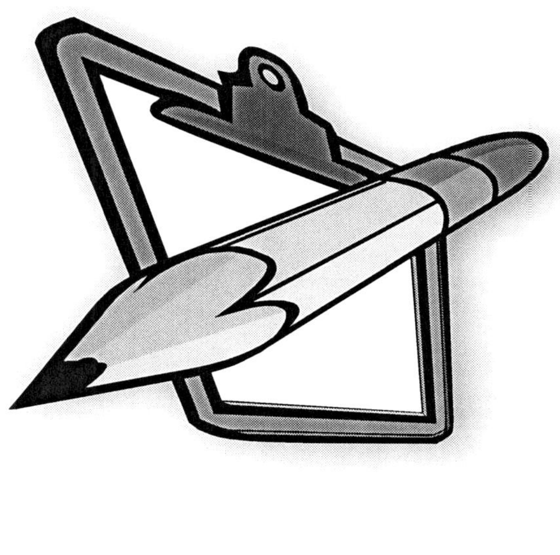

Nephrectomy

This morning the first case on the board for us is a simple nephrectomy. I will need to stop by the lab for any lab test results. On my way back I will check the chart for any known allergies, previous surgeries, or any other abnormalities. I will also speak to the patient, identify myself and ask if there are any questions that he might have regarding the surgery. I will make the team aware of anything extra that they may need. Once I get to the room, I will then check that the room is ready for the patient: Lab tests available, proper drugs available, x-rays hung, the case is open and all instrumentation is on hand and working properly, that anesthesia has spoken to the patient and we are ready to bring the patient to the room. The reason we are doing this surgery today is that the patient has severe hydronephrosis.

While the anesthesiologist is sedating the patient, I will assist with intubations by applying cricoid pressure (Sellecks procedure), if needed. With the patient asleep I can now place the Foley, and since this is an older male I will need to insert the catheter all the way, making sure that I have urine before inflating the balloon. The patient will be placed into lateral position. With the help of all the team members we will slowly move our patient to the edge of the table while anesthesia protects the head and airway and lift him on to his side. After placing pillows between the patient's legs and arms, and placing the lower leg flexed and the top leg straight, I can tape the patient into this position, break the table, and raise the kidney rest to fully expose the flank (kidney position).

Now I will need to place the bovie pad over a fatty location but still close to incision site. Also I should apply T.E.D. hose (antiembolic devise) for proper venous circulation. I will carefully place warm blankets over the arms and legs of my patient to prevent any hypothermia. The table strap is placed last to prevent the patient from a fall when the table is repositioned. With the site properly marked I can beep the surgeon and go scrub while the circulator does the prep. The Surgical technologist and I will drape the patient, and hook up the bovie and suction while the surgeon is scrubbing. A surgical timeout is done as to recognize the proper patient with their surgery and we can begin.

The incision we are going to use is subcostal flank. We only extend the incision as to get the proper exposure. I will use a small Richardson, as the superficial layers of tissue need to be exposed. With use of the bovie the Latissimus dorsi, external oblique, and internal oblique muscles are transected, (these muscles form Petit's triangle) next here we will identify the subcostal neurovascular bundle and I will gently retract with a vein retractor. Now we can place a self-retaining retractor and pack with moist laps. Next with use of Metzenbaums and Debakey pickups we can open the lumbodorsal fascia and partially split the transverse abdominus muscle.

Gerota's fascia is opened and with the help of a Deaver retractor we gain the needed exposure. We can expose the renal pedicle by using blunt dissection with cherry dissectors. Perinephric fat and the adrenal gland are dissected at the upper pole of the kidney with the use of the Mets and pickups. Once the surgeon has the ureter secured, clamped, and cut, we can double tie it. Next the kidney pedicles are exposed, double clamped, cut, and then tied (renal artery first, then renal vein). The technique I use is a one handed tie; this permits a secure knot with out having to get both hands into the wound. Also we use only nonabsorbable ties and double tie them due to the fact that these are large vessels and any bleeding here could be dangerous.

Last we are going to irrigate with warm saline and prepare to close. A drain will be placed now, as there might be blood or fluid build up. For my closure I will use a number 2 Vicryl runner with a figure eight interrupted stitch to close the peritoneum and fascia layers, and finish the closure with a 4-0 Vicryl subque. I will secure the drain with a 2-0 nylon. I will apply steri strips and mastasol the dressing will be fluffs and medipore tape.

The drapes are then removed and the table will return slowly to level so that the patient can be extubated. Care will be taken as we proceed to move the patient to his stretcher and transport to recovery. On my follow up I will advise the patient on dressing changes, showers, and basic movements. The expected outcome will be to return to normal activities in six to eight weeks.

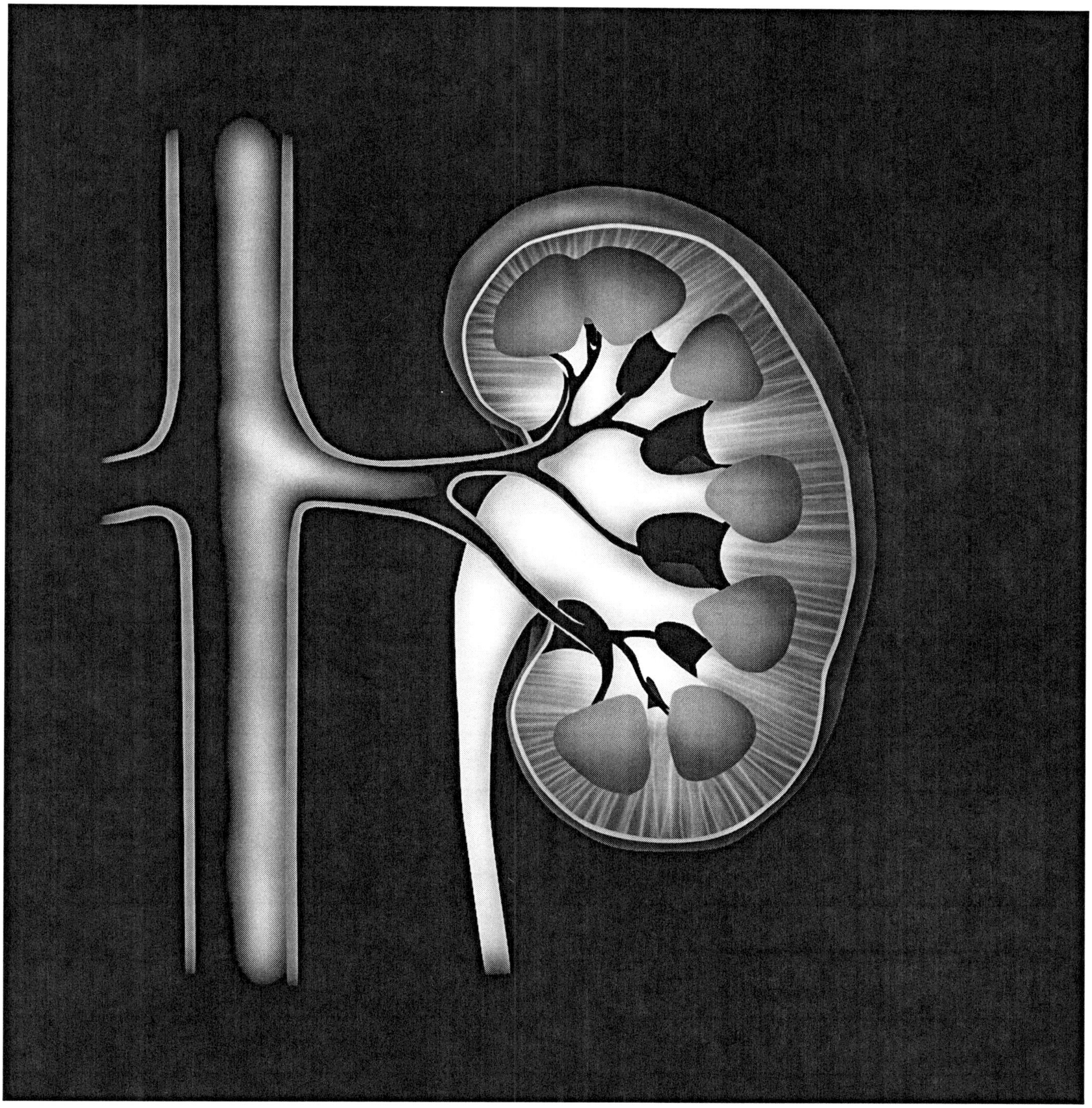

5.1 Kidney

Case study

In this section you will need to describe a nephrectomy in detail from your perspective (one that you have done in the past). List any differences in positioning, drugs used, or incision, etc.

1. During a nephrectomy, the preferred position of the patient is
   A. Lithotomy
   B. Prone
   C. Lateral
   D. Trendelenburg

2. A reason for performing a nephrectomy would be
   A. Hydronephrosis
   B. Tumor
   C. Calculus disease
   D. All the above

3. The incision we will make on a nephrectomy should be
   A. Pfannenstiel
   B. Mcburney
   C. Lower midline
   D. Flank

4. These three muscle groups form what triangle; Latissimus dorsi, external oblique, and internal oblique
   A. Petits
   B. Carrion
   C. Incisional
   D. Mcburney

5. The subcostal neurovascular bundle runs directly under which rib
   A. Tenth
   B. Eleventh
   C. Twelfth
   D. All ribs

6. The innermost layer of the kidney or true capsule is sometimes called
   A. Medulla
   B. Adrenal
   C. Gerota's
   D. Calyx

7. The paired organs that lie on the medial sides of the upper kidney are
   A. Medulla
   B. Adrenal glands
   C. Gerota's
   D. Calyx

8. The notched area located on the medial side of the kidney is known as the
   A. Renal
   B. Hilum
   C. Calyces
   D. Medulla

9. One of the most useful retractors for delicate vessels is the

    A. Gil-Vernet

    B. Love

    C. Hoen

    D. Satinsky

10. When the patient has a latex allergy, the catheter that is placed should be

    A. Foley

    B. Silicone

    C. Coude

    D. Robinson

11. Just as the kidneys are incased with peritoneum, the bladder is incased with

    A. Rugae

    B. Muscle

    C. Peritoneum

    D. Ligaments

12. Of the arteries that enter the pelvis, which two are not paired

    A. Sacral, rectal

    B. Sacral, iliac

    C. Iliac, rectal

    D. Iliac, ovarian

13. This artery supplies most of the pelvis with blood
    A. Sacral
    B. Iliac
    C. Rectal
    D. Ovarian

14. The ureters pass over the iliacs at this point
    A. L5- S1
    B. S1- S2
    C. L4- L5
    D. They do not pass over

15. Prenatally the umbilical arteries carry blood for reoxygenation, after birth this structure becomes the
    A. Artery
    B. Vein
    C. Umbilical ligament
    D. Umbilical fold

16. Occasionally an accessory renal artery will show on the ________ side of the aorta
    A. Right
    B. Left
    C. Either side
    D. Never occurs

17. The function of the suprarenal gland is to secrete

    A. Blood

    B. Urine

    C. Hormones

    D. Cretin

18. In about one in six hundred births an abnormality occurs that we call a _________ kidney

    A. Enlarged

    B. Undersized

    C. Drooped

    D. Horseshoe

19. Although uncommon, a kidney can appear on the

    A. Stomach

    B. Bladder

    C. Uterus

    D. Colon

20. The last resort of retrieving a kidney stone should be

    A. Lithotripsy

    B. Laser

    C. Basketing

    D. Percutaneous

21. Ureteric colic is pain of the ureter and is caused by

A. Distention

B. Stone

C. Calculus

D. All of the above

22. Large kidney stones may cause these symptoms

A. Obstruction

B. Bleeding

C. Infection

D. All of the above

23. One person in twenty-four hundred will have the kidney abnormality of

A. Horseshoe

B. Only one

C. Three

D. None of these

24. Accumulation of fluid in the peritoneal cavity is known as

A. Ascites

B. Cirrhosis

C. Cholecystokinin

D. Colic

25. A stone in the ureter will send pain radiating from the rib cage to the

    A. Groin

    B. Foot

    C. Pelvis

    D. Knee

26. The azygos vein which communicates with the renal veins drains this structure

    A. Heart

    B. Liver

    C. Gallbladder

    D. Abdomen wall

27. The right renal artery will pass ______to the vena cava

    A. Anterior

    B. Posterior

    C. Superior

    D. Inferior

28. The most lethal of inflammatory conditions is a bacterial invasion of the bloodstream or

    A. Heart disease

    B. Cancer

    C. Sepsis

    D. Diabetes

29. The arterial blood supply for the ureters are supplied from the________ artery

   A. Renal
   B. Superior vesical
   C. Superior uterine
   D. All of the above

30. Sometimes during a nephrectomy a rib is sacrificed, usually this one

   A. Seventh
   B. Tenth
   C. Twelfth
   D. First

31. The humpback shape of the kidney is designed for

   A. Urine flow
   B. Blood flow
   C. Placement
   D. No reason

32. The organ that we are careful to dissect around when removing the left kidney is the

   A. Liver
   B. Spleen
   C. Gallbladder
   D. Heart

33. The organ that we are careful to dissect around when removing the right kidney is the

A. Liver

B. Spleen

C. Gallbladder

D. Heart

34. Because of this organ, the right kidney will lie lower than the left

A. Spleen

B. Liver

C. Gallbladder

D. Heart

35. The term subphrenic refers to

A. Below kidney

B. Below liver

C. Under pancreas

D. Under diaphragm

36. The surgical removal of a small part or pie shape is

A. Wedge resection

B. Lobe resection

C. Pole resection

D. Biopsy

37. The original site of cancer that spreads or metastasis to other locations is called
    A. Original site
    B. Insitu
    C. Primary site
    D. Secondary

38. The tissue that surrounds a tumor is often described as
    A. Primary
    B. Secondary
    C. Marginal
    D. Metastatic

39. Cirrhosis caused by alcoholism is a term that is associated with this organ
    A. Kidney
    B. Liver
    C. Pancreas
    D. Gallbladder

40. The tissue that makes up organs such as the liver and the spleen is called
    A. Fibrous
    B. Connective
    C. Vascular
    D. Parenchyma

41. The conditions set into place that causes the spread of tumorous tissue is referred to as

A. Primary
B. Secondary
C. Metastatic
D. Insitu

42. When in the normal position or normal place, without having to disturb surrounding tissue is called

A. Primary
B. Secondary
C. Metastatic
D. Insitu

43. The muscle that controls the urinary bladder functions, uniformity, and capacity is

A. Detrusor
B. Trigone
C. Vesical
D. Pelvic

44. The ureters connect proximally to the renal pelvis and distally to the

A. Bladder
B. Kidney
C. Trigone
D. Urethra

45. Cystic and dysplastic lesions on the kidney are consistent with

A. Agenesis

B. Dysplasia

C. Polycystic

D. Pyelonephritis

46. We refer to multicystic kidneys as

A. Agenesis

B. Dysplasia

C. Polycystic

D. Pyelonephritis

47. Complete lack of urethral buds during fetus development and later causing infant death

A. Agenesis

B. Dysplasia

C. Polycystic

D. Pyelonephritis

48. The prostatic gland drains through about how many paired excretory ducts into the floor of the urethra

A. Six

B. Eight

C. Ten

D. Twelve

49. The muscular sheath that surrounds the ureter and extends into the trigone is the

    A. Hiatus

    B. Vesical

    C. Waldeyers

    D. Fossa

50. With cystitis the patient will present with these symptoms

    A. Dysuria, urgency

    B. Frequency, fever

    C. Back pain, hematuria

    D. All of the above

51. During a nephrectomy, the preferred position of the patient is Lateral

    A. True

    B. False

52. A tumor is one of the main reasons for performing a nephrectomy

    A. True

    B. False

53. A flank incision is the incision of choice for a nephrectomy

    A. True

    B. False

54. These three muscle groups form the Pettis triangle; Latissimus dorsi, external oblique, and internal oblique

    A. True

    B. False

55. The subcostal neurovascular bundle runs directly under all ribs

    A. True

    B. False

56. The innermost layer of the kidney or true capsule is sometimes called medulla

    A. True

    B. False

57. The paired organs that lie on the medial sides of the upper kidney are called medulla

    A. True

    B. False

58. The hilum is the notched area located on the medial side of the kidney

    A. True

    B. False

59. The Gil-Vernet is one of the most useful retractors for delicate vessels

    A. True

    B. False

60. Silicone catheters should be used when the patient has a latex allergy

    A. True

    B. False

61. Just as the kidneys are incased with peritoneum, the bladder also is incased with peritoneum

    A. True

    B. False

62. The sacral, rectal arteries are the only two arteries that enter the pelvis, which and are not paired

    A. True

    B. False

63. The rectal artery supplies most of the pelvis with blood

    A. True

    B. False

64. The ureters pass over the iliacs at L4- L5

    A. True

    B. False

65. Prenatally the umbilical arteries carry blood for reoxygenation, after birth this structure becomes the umbilical ligament

    A. True

    B. False

66. Occasionally an accessory renal artery will show on the left side of the aorta

    A. True

    B. False

67. The function of the suprarenal gland is to secrete urine

    A. True

    B. False

68. In about one in six hundred births an abnormality occurs that we call a horseshoe kidney

    A. True

    B. False

69. Although uncommon, a kidney can appear on the colon

    A. True

    B. False

70. The last resort of retrieving a kidney stone should be laser

    A. True

    B. False

71. Ureteric colic is pain of the ureter and is caused by trauma

    A. True

    B. False

72. Large kidney stones may cause bleeding, infection, and obstruction

    A. True

    B. False

73. One person in twenty-four hundred will have the kidney abnormality of one kidney

    A. True

    B. False

74. Accumulation of fluid in the peritoneal cavity is known as colic

    A. True

    B. False

75. A stone in the ureter will send pain radiating from the rib cage to the groin

    A. True

    B. False

76. The azygos vein which communicates with the renal veins drains the abdominal wall

    A. True

    B. False

77. The right renal artery will pass inferior to the vena cava

    A. True

    B. False

78. The most lethal of inflammatory conditions is a bacterial invasion of the bloodstream or sepsis

A. True

B. False

79. The arterial blood supply for the ureters are supplied from the azygos artery

A. True

B. False

80. Sometimes during a nephrectomy the second rib is sacrificed

A. True

B. False

81. The humpback shape of the kidney is designed for urine flow

A. True

B. False

82. The organ that we are careful to dissect around when removing the left kidney is the spleen

A. True

B. False

83. The organ that we are careful to dissect around when removing the right kidney is the liver

A. True

B. False

84. Because of the liver, the right kidney will lie lower than the left

A. True

B. False

85. The term subphrenic refers to below the liver

A. True

B. False

86. The surgical removal of a small part or pie shape is biopsy

A. True

B. False

87. The original site of cancer that spreads or metastasis to other locations is called insitu

A. True

B. False

88. The tissue that surrounds a tumor is often described as marginal

A. True

B. False

89. Cirrhosis caused by alcoholism is a term that is associated with the kidney

A. True

B. False

90. The tissue that makes up organs such as the liver and the spleen is extremely vascular

A. True

B. False

91. The conditions set into place that causes the spread of tumorous tissue is referred to as insitu

A. True

B. False

92. When things are located in the normal position or normal place; this is insitu

A. True

B. False

93. The muscle that controls the urinary bladder functions, uniformity, and capacity is the vesical

A. True

B. False

94. The ureters connect proximally to the renal pelvis and distally to the trigone

A. True

B. False

95. Cystic and dysplastic lesions on the kidney are consistent with polycystic disease

A. True

B. False

96. We refer to multicystic kidneys as dysplasia

    A. True

    B. False

97. Complete lack of urethral buds during fetus development and later causing infant death is agenesis

    A. True

    B. False

98. The prostatic gland drains through about twelve paired excretory ducts into the floor of the urethra

    A. True

    B. False

99. The muscular sheath that surrounds the ureter and extends into the trigone is called Waldeyers

    A. True

    B. False

100. With cystitis the patient will present with symptoms such as pain, bleeding, and fever

    A. True

    B. False

1 A ❑ B ❑ C ❑ D ❑
2 A ❑ B ❑ C ❑ D ❑
3 A ❑ B ❑ C ❑ D ❑
4 A ❑ B ❑ C ❑ D ❑
5 A ❑ B ❑ C ❑ D ❑
6 A ❑ B ❑ C ❑ D ❑
7 A ❑ B ❑ C ❑ D ❑
8 A ❑ B ❑ C ❑ D ❑
9 A ❑ B ❑ C ❑ D ❑
10 A ❑ B ❑ C ❑ D ❑
11 A ❑ B ❑ C ❑ D ❑
12 A ❑ B ❑ C ❑ D ❑
13 A ❑ B ❑ C ❑ D ❑
14 A ❑ B ❑ C ❑ D ❑
15 A ❑ B ❑ C ❑ D ❑
16 A ❑ B ❑ C ❑ D ❑
17 A ❑ B ❑ C ❑ D ❑
18 A ❑ B ❑ C ❑ D ❑
19 A ❑ B ❑ C ❑ D ❑
20 A ❑ B ❑ C ❑ D ❑
21 A ❑ B ❑ C ❑ D ❑
22 A ❑ B ❑ C ❑ D ❑
23 A ❑ B ❑ C ❑ D ❑
24 A ❑ B ❑ C ❑ D ❑
25 A ❑ B ❑ C ❑ D ❑
26 A ❑ B ❑ C ❑ D ❑
27 A ❑ B ❑ C ❑ D ❑
28 A ❑ B ❑ C ❑ D ❑
29 A ❑ B ❑ C ❑ D ❑
30 A ❑ B ❑ C ❑ D ❑
31 A ❑ B ❑ C ❑ D ❑
32 A ❑ B ❑ C ❑ D ❑
33 A ❑ B ❑ C ❑ D ❑
34 A ❑ B ❑ C ❑ D ❑
35 A ❑ B ❑ C ❑ D ❑
36 A ❑ B ❑ C ❑ D ❑
37 A ❑ B ❑ C ❑ D ❑
38 A ❑ B ❑ C ❑ D ❑
39 A ❑ B ❑ C ❑ D ❑
40 A ❑ B ❑ C ❑ D ❑
41 A ❑ B ❑ C ❑ D ❑
42 A ❑ B ❑ C ❑ D ❑
43 A ❑ B ❑ C ❑ D ❑
44 A ❑ B ❑ C ❑ D ❑
45 A ❑ B ❑ C ❑ D ❑
46 A ❑ B ❑ C ❑ D ❑
47 A ❑ B ❑ C ❑ D ❑
48 A ❑ B ❑ C ❑ D ❑
49 A ❑ B ❑ C ❑ D ❑
50 A ❑ B ❑ C ❑ D ❑
51 A ❑ B ❑ C ❑ D ❑
52 A ❑ B ❑ C ❑ D ❑
53 A ❑ B ❑ C ❑ D ❑
54 A ❑ B ❑ C ❑ D ❑
55 A ❑ B ❑ C ❑ D ❑
56 A ❑ B ❑ C ❑ D ❑
57 A ❑ B ❑ C ❑ D ❑
58 A ❑ B ❑ C ❑ D ❑
59 A ❑ B ❑ C ❑ D ❑
60 A ❑ B ❑ C ❑ D ❑
61 A ❑ B ❑ C ❑ D ❑
62 A ❑ B ❑ C ❑ D ❑
63 A ❑ B ❑ C ❑ D ❑
64 A ❑ B ❑ C ❑ D ❑
65 A ❑ B ❑ C ❑ D ❑
66 A ❑ B ❑ C ❑ D ❑
67 A ❑ B ❑ C ❑ D ❑
68 A ❑ B ❑ C ❑ D ❑
69 A ❑ B ❑ C ❑ D ❑
70 A ❑ B ❑ C ❑ D ❑
71 A ❑ B ❑ C ❑ D ❑
72 A ❑ B ❑ C ❑ D ❑
73 A ❑ B ❑ C ❑ D ❑
74 A ❑ B ❑ C ❑ D ❑
75 A ❑ B ❑ C ❑ D ❑
76 A ❑ B ❑ C ❑ D ❑
77 A ❑ B ❑ C ❑ D ❑
78 A ❑ B ❑ C ❑ D ❑
79 A ❑ B ❑ C ❑ D ❑
80 A ❑ B ❑ C ❑ D ❑
81 A ❑ B ❑ C ❑ D ❑
82 A ❑ B ❑ C ❑ D ❑
83 A ❑ B ❑ C ❑ D ❑
84 A ❑ B ❑ C ❑ D ❑
85 A ❑ B ❑ C ❑ D ❑
86 A ❑ B ❑ C ❑ D ❑
87 A ❑ B ❑ C ❑ D ❑
88 A ❑ B ❑ C ❑ D ❑
89 A ❑ B ❑ C ❑ D ❑
90 A ❑ B ❑ C ❑ D ❑
91 A ❑ B ❑ C ❑ D ❑
92 A ❑ B ❑ C ❑ D ❑
93 A ❑ B ❑ C ❑ D ❑
94 A ❑ B ❑ C ❑ D ❑
95 A ❑ B ❑ C ❑ D ❑
96 A ❑ B ❑ C ❑ D ❑
97 A ❑ B ❑ C ❑ D ❑
98 A ❑ B ❑ C ❑ D ❑
99 A ❑ B ❑ C ❑ D ❑
100 A ❑ B ❑ C ❑ D ❑

CH5 Nephrectomy

Multiple Choice

1. C	11. C	21. D	31. D	41. C
2. D	12. A	22. D	32. B	42. D
3. D	13. B	23. B	33. A	43. A
4. D	14. A	24. A	34. B	44. C
5. D	15. C	25. A	35. B	45. C
6. C	16. B	26. D	36. A	46. B
7. B	17. C	27. B	37. C	47. A
8. B	18. D	28. C	38. C	48. D
9. A	19. C	29. D	39. B	49. C
10. B	20. D	30. A	40. D	50. D

True / False

51. True	61. True	71.Distention, Stone, Calculus	81. No reason	91. Metastatic
52. True	62. True	72. True	82. True	92. True
53. True	63. Iliac	73. True	83. True	93. Detrusor
54. Mcburney	64. L5, S1	74. Ascites	84. True	94. True
55. True	65. True	75. True	85. True	95. True
56. Gerota's	66. True	76. True	86. Wedge resection	96. True
57. Adrenal gland	67. Hormones	77. Posterior	87. Primary site	97. True
58. True	68. True	78. True	88. True	98. True
59. True	69. Bladder	79. Renal, Superior vessel, Superior uterine	89. Liver	99. True
60. True	70.Percutaneous	80. Twelfth	90. True	100. True

CHAPTER 6

AORTIC ANEURYSM CASE STUDY

Aortic Aneurysm

This morning the first case on the board for us is an abdominal aortic aneurysm. I will need to stop by the lab for any test results of which I should be aware. On my way back I will check the chart for any known allergies, previous surgeries, or any other abnormalities. I will also speak to the patient, identify myself and ask if there are any questions that she might have regarding the surgery. I will make the team aware of anything extra that they may need. Once I get to the room, I will check that the room is ready for the patient: Lab tests available, proper drugs available, the case is open and all instrumentation is on hand and working properly, that anesthesia has spoken to the patient and we are ready to bring the patient to the room. The reason we are doing this surgery today is the patient has severe abdominal and back pain with a pulsatile abdominal mass. The patient is at a very high risk of rupture because of the addition of hypertension and obesity.

While the anesthesiologist is sedating the patient I will assist with intubations by applying cricoid pressure (Sellecks procedure) if needed. With the patient asleep I can now place the Foley, and since this is a female I will only need to insert the catheter about four inches, making sure that I have urine before inflating balloon. The patient is placed into supine position. With the help of all the team members we will slowly move our patient to the edge of the table while anesthesia protects the head and airway and then move.

Now I will need to place the bovie pad over a fatty location but still close to incision site. Also I should apply T.E.D. hose (antiembolic devise) for proper venous circulation. I will carefully place warm blankets over the arms and legs of my patient to prevent any hypothermia. The table strap is placed last to prevent the patient from a fall when the table is repositioned. Since the head of the table will be dropped, the patient's knees should be over a break in the bed so the legs can be lowered also. With the site properly marked I can beep the surgeon and go scrub while the circulator does the prep. The Surgical technologist and I will drape the patient and hook up bovie and suction while the surgeon is scrubbing. A surgical timeout is done as to recognize the proper patient with their surgery and we can begin.

The incision we are going to use is midline. We only extend the incision as to get the proper exposure. I will use a small Richardson, as the superficial layers of tissue need to be exposed. With use of the bovie the fascia layers are opened horizontally along the linea alba next we will very carefully open the peritoneum vertically from xiphoid process to pubis. I will then place moistened lap sponges at the wound edges under the Bookwalter retractor blades to avert any nerve damage.

Now we displace the bowel over the wound to the right and cover with warm moistened towels to gain maximum exposure. Now the posterior peritoneum is incised and extended over the common iliac arteries. The mesenteries can be freed and vessel loops used to avert any back bleeding. We are now tunneling under the iliacs for our eventual clamping of these vessels. At the proximal end of the aorta we have cleared for placement of our aortic clamp just below the renal arteries. An eight mm diameter bifurcated graft is selected and given to the technologist for pre-clotting. The heparin can be given now intravenously as we are about to clamp the common iliacs. The clock is started and we will now clamp the aorta above the aneurysm.

We will open the aorta longitudinally along the anterior lateral aspect and stop just short of the bifurcation of the iliacs. All thrombus material is removed from the vessel. A horizontal incision is made at the proximal end of the aorta. We will now irrigate with warm saline as to not disrupt the core temperature. A 4-0 Prolene will be used to anastomosis the proximal end of the graft. With this completed we can check our suture line for leaks by applying another clamp just distal to our anastomosis and releasing the proximal one. Now that this part is finished we can work on the bifurcation end. We will cut the graft to the proper length and suture the graft with 5-0 Prolene to the common iliacs. With this completed we will release our proximal clamp and check for leaks on this end. All finished with the graft, we can now release the distal clamps, close the aorta sac with a 2-0 vicryl over the graft so that the bowel will not be harmed by the constant motion of the aorta, and irrigate once more with warm saline.

For my closure I will use a number 2 Vicryl runner to close the fascia layers and finish the closure with a skin gun. For the dressing I will apply fluffs, abd's and medipore tape. The drapes are then removed and the table will be returned slowly to level so that the patient can be extubated. Care will be taken as we proceed to move the patient to her stretcher and transported to CVCU. On my follow-up I will advise the patient on dressing changes, showers, and basic movements. The expected outcome will be to return to normal activities.

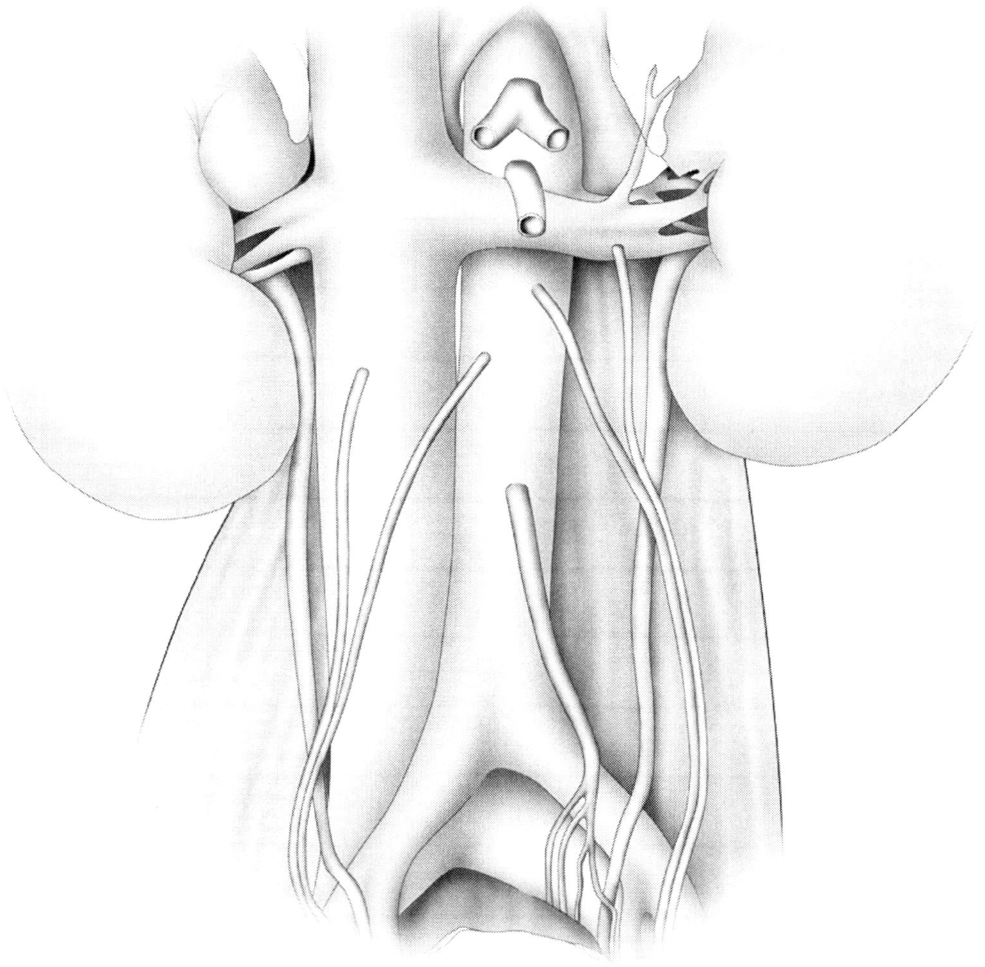

6.1 Aorta

Case study

In this section you will need to describe an abdominal aortic aneurism in detail from your perspective (one that you have done in the past). List any differences in positioning, drugs used, or incision, etc.

1. The tourniquets that are fashioned from red rubber tubing and umbilical tape are called
   A. Rumel
   B. Shunt tape
   C. Loops
   D. Clamp

2. The drug used to reverse the effects of Heparin is
   A. Mannitol
   B. Sodium bicarbonate
   C. Papaverine
   D. Protamine

3. The drug used to dilate peripheral blood vessels
   A. Mannitol
   B. Sodium bicarbonate
   C. Papaverine
   D. Protamine

4. The painful feeling of cramping or dull ache in the abdomen is known as
   A. Stenosis
   B. Thrombus
   C. Claudication
   D. Embolus

5. The narrowing or stricture of a blood vessel is known as

   A. Stenosis

   B. Thrombus

   C. Claudication

   D. Embolus

6. The instruments used for vascular clamping are

   A. DeBakey

   B. Lambert-Kay

   C. Cooley

   D. All of the above

7. A typical suture gauge for anastomosis of the aorta is

   A. 3-0, 4-0

   B. 4-0, 5-0

   C. 5-0, 6-0

   D. 6-0, 7-0

8. The aorta passes through the thoracic diaphragm from a structure known as

   A. Foramen

   B. Hiatus

   C. Gap

   D. Crura

9. Where does the vena cava lie in perspective to the aorta

    A. Anterior

    B. Posterior

    C. Anterolateral

    D. Posterior lateral

10. The aorta divides distally into how many vessels

    A. Two

    B. Four

    C. Three

    D. One

11. The left renal vein comes into contact with the aorta

    A. Dorsally

    B. Does not

    C. Anterior

    D. Posterior

12. The right renal vein comes into contact with the aorta

    A. Dorsally

    B. Does not

    C. Anterior

    D. Posterior

13. The common iliac vein comes into contact with the aorta

    A. Dorsally

    B. Does not

    C. Anterior

    D. Posterior

14. The splenic artery comes off this trunk

    A. Aorta

    B. Hepatic

    C. Mesenteric

    D. Celiac

15. The gastroduodenal artery comes off this trunk

    A. Aorta

    B. Hepatic

    C. Mesenteric

    D. Celiac

16. The pulsations of the abdominal aorta can be felt at this vertebral level

    A. L1-L2

    B. L2-L3

    C. L4

    D. L5

17. The left renal artery comes into contact with the vena cava

    A. Medially

    B. Does not

    C. Anterior

    D. Posterior

18. The right renal artery comes into contact with the vena cava

    A. Medially

    B. Does not

    C. Anterior

    D. Posterior

19. The superior mesenteric leaves the aorta between the renal arteries and the

    A. Celiac trunk

    B. Inferior mesenteric

    C. Iliacs

    D. Hepatics

20. The gonadal arteries arise from the aorta below this point

    A. Celiac trunk

    B. Renals

    C. Iliacs

    D. Hepatics

21. Most common size of an aortic graft used is

    A. 8mm

    B. 9mm

    C. 10mm

    D. 11mm

22. The largest vein in the body is

    A. Hepatic

    B. Vena cava

    C. Common iliacs

    D. Femoral

23. The right azygos vein originates from this venous structure

    A. Subcostal

    B. Lumbar

    C. Mesenteric

    D. A and B

24. During a laminectomy which vessel is at risk of injury

    A. Vena cava

    B. Iliacs

    C. Aorta

    D. A and B

25. The cisterna chyli is part of this system
    - A. Venous
    - B. Arterial
    - C. Lymphatic
    - D. Endocrine

26. The cisterna chyli's main duct
    - A. Thoracic
    - B. Lumbar
    - C. Sacral
    - D. Femoral

27. In an emergency the best incision for an abdominal aortic aneurism is
    - A. Pfannenstiel
    - B. Trans abdominal
    - C. Midline
    - D. Kockers

28. Portal hypertension is a classic complication of this disease
    - A. Cystitis
    - B. Cirrhosis
    - C. Hemorrhoids
    - D. Splenomegaly

29. A classic complication of portal hypertension is

    A. Cystitis

    B. Cirrhosis

    C. Hemorrhoids

    D. Splenomegaly

30. Caput medusae is characterized by

    A. Varicose veins

    B. Cirrhosis

    C. Hemorrhoids

    D. Fibroids

31. The small intestine and most of the colon are drained by this vein

    A. Superior mesenteric

    B. Splenic

    C. Inferior mesenteric

    D. Gastroomental

32. The descending colon and rectum are drained by this vein

    A. Superior mesenteric

    B. Splenic

    C. Inferior mesenteric

    D. Gastroomental

33. The stomach is drained of blood by this vein

    A. Superior mesenteric

    B. Splenic

    C. Inferior mesenteric

    D. Gastroomental

34. The inferior mesenteric vein will drain directly to the

    A. Superior mesenteric

    B. Splenic

    C. Hepatic portal

    D. Common iliac

35. The testis and ovaries drain into this vein

    A. Suprarenal

    B. Gonadal

    C. Phrenic

    D. None of the above

36. The diaphragm will drain into this vein

    A. Suprarenal

    B. Gonadal

    C. Phrenic

    D. None of the above

37. The pelvis and its viscera will drain into this vein

    A. External iliac

    B. Internal iliac

    C. Gonadal

    D. Ascending lumbar

38. The lower limbs are drained by this vein

    A. External iliac

    B. Internal iliac

    C. Gonadal

    D. Ascending lumbar

39. The right posterior thoracic wall is drained by this vein

    A. Hemiazygos

    B. Azygos

    C. Brachiocephalic

    D. Accessory azygos

40. The left posterior thoracic wall is drained by this vein

    A. Hemiazygos

    B. Azygos

    C. Brachiocephalic

    D. Accessory azygos

41. The artery that supplies the knee and leg would be the

    A. Tibial

    B. Fibular

    C. Digital

    D. Plantar

42. The artery that supplies the calf and ankle would be the

    A. Tibial

    B. Fibular

    C. Digital

    D. Plantar

43. The artery that supplies the thigh and knee would be the

    A. Popliteal

    B. Fibular

    C. Femoral

    D. Tibial

44. The drug most likely to trigger malignant hyperthermia would be

    A. Nitrous oxide

    B. Halothane

    C. Dantrolene

    D. Succinycholine

45. The drug of choice given for malignant hyperthermia

    A. Nitrous oxide

    B. Halothane

    C. Dantrolene

    D. Succinylcholine

46. The safest way to put a patient to sleep that has a susceptible risk for malignant hyperthermia is

    A. Halothane

    B. Nitrous oxide

    C. Succinylcholine

    D. None are safe

47. Injury to a patient's brachial plexus will result from hyper- extending the arm

    A. 90 degrees

    B. 45 degrees

    C. 30 degrees

    D. Will not cause harm

48. A loss of excessive bicarbonate in the blood system will result in

    A. Metabolic alkalosis

    B. Metabolic acidosis

    C. Hypomagnesaemia

    D. Hypokalemia

49. An increase of bicarbonate in the blood system will result in

A. Metabolic alkalosis

B. Metabolic acidosis

C. Hypomagnesaemia

D. Hypokalemia

50. Low blood potassium disorder is referred to as

A. Metabolic alkalosis

B. Metabolic acidosis

C. Hypomagnesaemia

D. Hypokalemia

51. The tourniquets that are fashioned from red rubber tubing and umbilical tape are called loops

A. True

B. False

52. The drug used to reverse the effects of Heparin is Protamine

A. True

B. False

53. The drug used to dilate peripheral blood vessels is Papaverine

A. True

B. False

54. The painful feeling of cramping or dull ache in the abdomen is known as Claudication

A. True

B. False

55. The narrowing or stricture of a blood vessel is known as stenosis

A. True

B. False

56. DeBakey, Lambert-Kay, and Cooley are all instruments used for vascular clamping

A. True

B. False

57. A typical suture gauge for anastomosis of the aorta is 10-0

A. True

B. False

58. The aorta passes through the thoracic diaphragm from a structure known as Crura

A. True

B. False

59. The vena cava lies posterior lateral in perspective to the aorta

A. True

B. False

60. The aorta divides distally into two vessels

    A. True

    B. False

61. The left renal vein comes into contact with the aorta on the anterior aspect

    A. True

    B. False

62. The right renal vein comes into contact with the aorta anterior aspect

    A. True

    B. False

63. The common iliac vein comes into contact with the aorta on the posterior aspect

    A. True

    B. False

64. The splenic artery comes off the celiac trunk

    A. True

    B. False

65. The gastroduodenal artery comes off the celiac trunk

    A. True

    B. False

66. The pulsations of the abdominal aorta can be felt at the L4 vertebral level

    A. True

    B. False

67. The left renal artery comes into contact with the vena cava on the posterior aspect

A. True

B. False

68. The right renal artery comes into contact with the vena cava on the posterior aspect

A. True

B. False

69. The superior mesenteric leaves the aorta between the renal arteries and the celiac trunk

A. True

B. False

70. The gonadal arteries arise from the aorta below the renals

A. True

B. False

71. Most common size of an aortic graft used is 6mm

A. True

B. False

72. The vena cava is the largest vein in the body

A. True

B. False

73. The right azygos vein originates from the subcostal and the lumbar veins

A. True

B. False

74. During a laminectomy the aorta and the vena cava vessels are at risk of injury

A. True

B. False

75. The cisterna chyli is part of the lymphatic system

A. True

B. False

76. The cisterna chyli's main duct is thoracic

A. True

B. False

77. In an emergency the best incision for an abdominal aortic aneurysm is a midline laparotomy

A. True

B. False

78. Portal hypertension is a classic complication of splenomegaly disease

A. True

B. False

79. A classic complication of portal hypertension is hemorrhoids

    A. True

    B. False

80. Caput medusae is characterized by varicose veins

    A. True

    B. False

81. The small intestine and most of the colon are drained by the superior mesenteric vein

    A. True

    B. False

82. The descending colon and rectum are drained by the superior mesenteric vein

    A. True

    B. False

83. The stomach is drained of blood by the gastroomental vein

    A. True

    B. False

84. The inferior mesenteric vein will drain directly to the hepatic portal

    A. True

    B. False

85. The testis and ovaries drain into the gonadal vein

    A. True

    B. False

86. The diaphragm will drain into the suprarenal vein

    A. True

    B. False

87. The pelvis and its viscera will drain into the internal iliac vein

    A. True

    B. False

88. The lower limbs are drained by the internal iliac vein

    A. True

    B. False

89. The right posterior thoracic wall is drained by the azygos vein

    A. True

    B. False

90. The left posterior thoracic wall is drained by the hemiazygos vein

    A. True

    B. False

91. The tibial artery supplies the knee and leg

    A. True

    B. False

92. The tibial artery supplies the calf and ankle

A. True

B. False

93. The femoral artery supplies the thigh and knee

A. True

B. False

94. The drug most likely to trigger malignant hyperthermia would be Succinycholine

A. True

B. False

95. The drug of choice given for malignant hyperthermia is Dantrolene

A. True

B. False

96. The safest way to put a patient to sleep that has a susceptible risk for malignant hyperthermia is with Succinycholine

A. True

B. False

97. Injury to a patient's brachial plexus will result from hyper extending the arm 90 degrees

A. True

B. False

98. A loss of excessive bicarbonate in the blood system will result in metabolic acidosis

    A. True

    B. False

99. An increase of bicarbonate in the blood system will result in metabolic acidosis

    A. True

    B. False

100. Low blood potassium disorder is referred to as hypokalemia

    A. True

    B. False

1	A ❑	B ❑	C ❑	D ❑
2	A ❑	B ❑	C ❑	D ❑
3	A ❑	B ❑	C ❑	D ❑
4	A ❑	B ❑	C ❑	D ❑
5	A ❑	B ❑	C ❑	D ❑
6	A ❑	B ❑	C ❑	D ❑
7	A ❑	B ❑	C ❑	D ❑
8	A ❑	B ❑	C ❑	D ❑
9	A ❑	B ❑	C ❑	D ❑
10	A ❑	B ❑	C ❑	D ❑
11	A ❑	B ❑	C ❑	D ❑
12	A ❑	B ❑	C ❑	D ❑
13	A ❑	B ❑	C ❑	D ❑
14	A ❑	B ❑	C ❑	D ❑
15	A ❑	B ❑	C ❑	D ❑
16	A ❑	B ❑	C ❑	D ❑
17	A ❑	B ❑	C ❑	D ❑
18	A ❑	B ❑	C ❑	D ❑
19	A ❑	B ❑	C ❑	D ❑
20	A ❑	B ❑	C ❑	D ❑
21	A ❑	B ❑	C ❑	D ❑
22	A ❑	B ❑	C ❑	D ❑
23	A ❑	B ❑	C ❑	D ❑
24	A ❑	B ❑	C ❑	D ❑
25	A ❑	B ❑	C ❑	D ❑
26	A ❑	B ❑	C ❑	D ❑
27	A ❑	B ❑	C ❑	D ❑
28	A ❑	B ❑	C ❑	D ❑
29	A ❑	B ❑	C ❑	D ❑
30	A ❑	B ❑	C ❑	D ❑
31	A ❑	B ❑	C ❑	D ❑
32	A ❑	B ❑	C ❑	D ❑
33	A ❑	B ❑	C ❑	D ❑
34	A ❑	B ❑	C ❑	D ❑
35	A ❑	B ❑	C ❑	D ❑
36	A ❑	B ❑	C ❑	D ❑
37	A ❑	B ❑	C ❑	D ❑
38	A ❑	B ❑	C ❑	D ❑
39	A ❑	B ❑	C ❑	D ❑
40	A ❑	B ❑	C ❑	D ❑
41	A ❑	B ❑	C ❑	D ❑
42	A ❑	B ❑	C ❑	D ❑
43	A ❑	B ❑	C ❑	D ❑
44	A ❑	B ❑	C ❑	D ❑
45	A ❑	B ❑	C ❑	D ❑
46	A ❑	B ❑	C ❑	D ❑
47	A ❑	B ❑	C ❑	D ❑
48	A ❑	B ❑	C ❑	D ❑
49	A ❑	B ❑	C ❑	D ❑
50	A ❑	B ❑	C ❑	D ❑
51	A ❑	B ❑	C ❑	D ❑
52	A ❑	B ❑	C ❑	D ❑
53	A ❑	B ❑	C ❑	D ❑
54	A ❑	B ❑	C ❑	D ❑
55	A ❑	B ❑	C ❑	D ❑
56	A ❑	B ❑	C ❑	D ❑
57	A ❑	B ❑	C ❑	D ❑
58	A ❑	B ❑	C ❑	D ❑
59	A ❑	B ❑	C ❑	D ❑
60	A ❑	B ❑	C ❑	D ❑
61	A ❑	B ❑	C ❑	D ❑
62	A ❑	B ❑	C ❑	D ❑
63	A ❑	B ❑	C ❑	D ❑
64	A ❑	B ❑	C ❑	D ❑
65	A ❑	B ❑	C ❑	D ❑
66	A ❑	B ❑	C ❑	D ❑
67	A ❑	B ❑	C ❑	D ❑
68	A ❑	B ❑	C ❑	D ❑
69	A ❑	B ❑	C ❑	D ❑
70	A ❑	B ❑	C ❑	D ❑
71	A ❑	B ❑	C ❑	D ❑
72	A ❑	B ❑	C ❑	D ❑
73	A ❑	B ❑	C ❑	D ❑
74	A ❑	B ❑	C ❑	D ❑
75	A ❑	B ❑	C ❑	D ❑
76	A ❑	B ❑	C ❑	D ❑
77	A ❑	B ❑	C ❑	D ❑
78	A ❑	B ❑	C ❑	D ❑
79	A ❑	B ❑	C ❑	D ❑
80	A ❑	B ❑	C ❑	D ❑
81	A ❑	B ❑	C ❑	D ❑
82	A ❑	B ❑	C ❑	D ❑
83	A ❑	B ❑	C ❑	D ❑
84	A ❑	B ❑	C ❑	D ❑
85	A ❑	B ❑	C ❑	D ❑
86	A ❑	B ❑	C ❑	D ❑
87	A ❑	B ❑	C ❑	D ❑
88	A ❑	B ❑	C ❑	D ❑
89	A ❑	B ❑	C ❑	D ❑
90	A ❑	B ❑	C ❑	D ❑
91	A ❑	B ❑	C ❑	D ❑
92	A ❑	B ❑	C ❑	D ❑
93	A ❑	B ❑	C ❑	D ❑
94	A ❑	B ❑	C ❑	D ❑
95	A ❑	B ❑	C ❑	D ❑
96	A ❑	B ❑	C ❑	D ❑
97	A ❑	B ❑	C ❑	D ❑
98	A ❑	B ❑	C ❑	D ❑
99	A ❑	B ❑	C ❑	D ❑
100	A ❑	B ❑	C ❑	D ❑

CH6 Aortic Aneurysm

Multiple Choice

1. A	11. C	21. A	31. A	41. A
2. D	12. B	22. B	32. C	42. B
3. C	13. D	23. D	33. D	43. C
4. C	14. D	24. D	34. C	44. D
5. A	15. D	25. C	35. B	45. C
6. D	16. C	26. A	36. D	46. D
7. A	17. B	27. C	37. B	47. A
8. B	18. D	28. B	38. A	48. B
9. D	19. A	29. C	39. B	49. A
10. B	20. B	30. A	40. A	50. B

True / False

51. Rumel	61. True	71. 8mm	81. True	91. True
52. True	62. Does not	72. True	82. True	92. Fibular
53. True	63. True	73. True	83. True	93. True
54. True	64. True	74. True	84. True	94. True
55. True	65. True	75. True	85. True	95. True
56. True	66. True	76. True	86. Phrenic	96. No safe way
57. 3-0,4-0	67. Does not	77. True	87. True	97. True
58. Hiatus	68. True	78. Cirrhosis	88. External Iliac	98. True
59. True	69. True	79. True	89. True	99. Metabolic acidosis
60. True	70. True	80. True	90. True	100. True

CHAPTER 7

HIP REVISION CASE STUDY

Hip Revision

This afternoon the first case on the board for us is a hip revision for degenerative arthritis. I will need to stop by x-ray and pick up the patient's folder containing x-rays, cat scans, MRIs and myelograms. On my way back, I will check the chart for any known allergies, previous surgeries, or any other abnormalities. I will also speak to the patient, identify myself and ask if there are any questions that he might have regarding the surgery. I will make the team aware of anything extra that they may need. Once I get to the room, I will check that the room is ready for the patient: x-rays hung, proper drugs available, the case is open and all instrumentation is on hand and working properly, that anesthesia has spoken to the patient and we are ready to bring the patient to the room.

While the anesthesiologist is sedating the patient, I will assist with intubations by applying cricoid pressure if needed. With the patient asleep I can now place the Foley, tape the eyes, and prepare the pads for the elbows as we prepare to flip the patient into lateral position. With the help of all the team members we will slowly move the patent to the edge, and roll the patient over on their side while anesthesia protects the head and airway. Once over I will place the arm on a pillow and an axillary roll under the patient's axillary. The head will be placed on a headrest and padding will then be placed under the knees and feet, the bottom leg will be flexed and the top one straight. Once the patient is fixed properly in the lateral frame, I will check genitals and arrange the Foley to hang at the head of the table. The affected leg will hang in a candy cane stirrup for prepping so I wrap the foot with webril so as not to cause any skin damage.

Now I will need to place the bovie pad over a fatty location but still close to incision site. I will carefully place warm blankets over the arms and legs of my patient to prevent hypothermia. The table strap is placed last to prevent the patient from a fall when the table is repositioned. With the site properly marked I can beep the surgeon and go scrub while the circulator does the prep. A surgical timeout is done as to recognize the proper patient with their surgery and we can begin.

We will make our incision anterior laterally about 20cm long on the affected hip and place small Gelpi self-retaining retractors. With the scalpel along with the bovie we will dissect down to the iliotibial band and retract the tensor fasciae latae medially and the gluteus medius laterally. While the surgeon is doing this I will use the Hibbs retractors to gain better exposure.

Now that we have maximum exposure we place a Charnley hip retractor in the wound. The proximal femoral shaft is now exposed. With the help of a bone hook I will now dislocate the hip. By using leverage I can pop the head out of the cup and I will place the knee firmly against my abdomen and let the leg hang freely in front of me. First the head and neck can be cut loose with our Stryker saw just above the trochanteric crest. Now we can begin our acetabular reaming gradually working from small to large and checking as we go along for bleeding bone and working for the largest cup size that the hip will accept. With this done and our cup and trial liner in we can start reaming and broaching the femoral canal. On the femur we want the largest implant that the canal will accept so as to not have any play in the canal.

Now that are implants are all in I will reduce the hip and the surgeon can press down with a head pusher. I will run the leg through a range of motion to check for a loose fit or dislocation; also we will check leg length against the opposite knee and foot. We will be sure that the laps are removed at the end of the case when the lap count is initialized. Last we are going to irrigate using our pulse-vac with bacitracin in warm saline and prepare to close.

The surgeon will probably leave now and leave me with placement of the drains and closure of the wound. For my closure I will use interrupted number 2 vicryl for the capsule and 0 vicryl on the next facial layers, then switch to 2-0 vicryl for the final dermal layer, staples for the skin and now we are ready for the dressing which will consist of Neosporin, Adaptec, Kerlex fluffs, ABD pads, and Medipore tape. We will connect the drain tubes and tape those down to the patient. Now we should be ready to carefully move the patient back on to his stretcher for extubation. Depending on the reason for the hip revision the outcome should be favorable with the patient able to return to full weight bearing in four to six weeks.

Case study

In this section you will need to describe a hip revision in detail from your perspective (one that you have done in the past). List any differences in positioning, drugs used, incision, etc.

1. A ROM test is generally used for this procedure
   A. Hip replacement
   B. Shoulder replacement
   C. Knee replacement
   D. All the above

2. The most common indication for a hip replacement is
   A. Rheumatoid arthritis
   B. Congenital dislocation
   C. Degenerative arthritis
   D. Painful hip

3. When dislocating a hip joint how many people should pry on the femur
   A. One
   B. Two
   C. Three
   D. No one

4. Which hip surgery will need a abduction pillow
   A. Anterior lateral
   B. Posterior lateral
   C. Supra lateral
   D. None of these

5. During a hip replacement this bone is in danger of fracture

    A. Femur

    B. Acetabulum

    C. Pubic ramus

    D. All the above

6. During a hip replacement this nerve is in danger of damage

    A. Sciatic

    B. Femoral

    C. Peroneal

    D. All the above

7. The largest bone in the body is the

    A. Femur

    B. Acetabulum

    C. Skull

    D. Iliac

8. The largest nerve in the body is the

    A. Femoral

    B. Sciatic

    C. Peroneal

    D. Tibial

9. The lumbar plexus includes L1-L4 what does the sacral plexus encompass

   A. S1-S3

   B. S1-S5

   C. L4-S3

   D. L1-S5

10. The femoral or sciatic nerve can be damaged during surgery by

    A. Position of extremity

    B. Instrument trauma

    C. Surgical trauma

    D. All the above

11. If the sciatic nerve is damaged during surgery it will take how long to recover from the trauma?

    A. One month

    B. One year

    C. Two weeks

    D. Ten years

12. Which nerve is at risk when the assistant is retracting the iliopsoas muscle and tendon

    A. Sciatic

    B. Femoral

    C. Peroneal

    D. Tibial

13. Which nerve is at risk during a routine hip posterior capsulotomy

A. Sciatic

B. Femoral

C. Peroneal

D. Tibial

14. The nerve most likely to be injured when dropping the foot down as during a hip procedure is

A. Sciatic

B. Femoral

C. Peroneal

D. Tibial

15. The sciatic nerve is easily identified by this surrounding tissue

A. Edemas

B. Fatty

C. Vascular

D. Scar

16. When the assistant takes measures to avoid injury to this nerve they will also be protecting the deep femoral artery and vein

A. Sciatic

B. Femoral

C. Peroneal

D. Tibial

17. When retracting in the area of a nerve, vein, or artery one should use this retractor
    - A. Sharp
    - B. Blunt
    - C. Self-retaining
    - D. Gelpi

18. Removal of the femoral head without replacement is termed
    - A. Arthronecrosis
    - B. Arthroplasty
    - C. Head removal
    - D. Girdlestone

19. This is a risk factor for increased rate of infection in an arthroplasty
    - A. Older age
    - B. Obesity
    - C. Diabetes
    - D. All of the above

20. This factor should cancel an operation that includes implants
    - A. Steroids
    - B. Arthritis
    - C. Urinary infection
    - D. Hematoma

21. Antibiotics used in cement will leach out locally for a certain amount time

    A. One day

    B. Three days

    C. Three weeks

    D. Six months

22. Organisms that infect joint replacement are usually

    A. Gram positive

    B. Gram negative

    C. Cloudy

    D. All of the above

23. When one uses a Charley incision the hip will be dislocated

    A. Anterior

    B. Posterior

    C. A or B

    D. None of the above

24. This disorder would warrant a hip replacement

    A. Arthritis

    B. Dislocation

    C. Girdlestone

    D. All of the above

25. For a younger person that is getting a hip replacement this implant is recommended

A. Metal on metal

B. Metal on polyethylene

C. Ceramic

D. Stainless steel

26. The weakest link in a hip procedure is

A. Bone

B. Cement

C. Implant

D. Poly

27. The actual load that a metal can withstand without bending or breaking is called

A. Fatigue limit

B. Fatigue strength

C. Yield strength

D. Toughness

28. The safest tourniquet to use for an extremity is

A. Esmarch

B. Martin

C. Pneumatic

D. Sheet

29. An extremity should be held up for how long to express the blood?

    A. Five minuets

    B. Two minuets

    C. Thirty seconds

    D. Forty five seconds

30. An adult arm tourniquet should be no more than ______ mm Hg

    A. 300

    B. 500

    C. 25

    D. 6

31. An adult thigh tourniquet should be no more than______ mm Hg

    A. 300

    B. 500

    C. 25

    D. 6

32. A safe time limit on an arm tourniquet normally should be

    A. Thirty minuets

    B. One hour

    C. One and half hour

    D. Two hours

33. A safe time limit on a thigh tourniquet normally should be

    A. Thirty minuets

    B. One hour

    C. One and half hour

    D. Two hours

34. The fibula articulates mainly with this bone

    A. Femur

    B. Tibia

    C. Patella

    D. Calcaneus

35. One of these bones is not in the foot

    A. Talus

    B. Cuboid

    C. Navicular

    D. Capitate

36. The turning of the foot outward is termed

    A. Protraction

    B. Subtraction

    C. Eversion

    D. Inversion

37. The turning of the foot inward is termed

    A. Protraction

    B. Subtraction

    C. Eversion

    D. Inversion

38. The greater sciatic notch is located on which surface of the ilium?

    A. Anterior

    B. Posterior

    C. Lateral

    D. Medial

39. The obturator foramen is located on which surface of the ilium?

    A. Isn't on ilium

    B. Posterior

    C. Lateral

    D. Medial

40. The femoral epicondyles are located on which condyle

    A. Anterior

    B. Both condyles

    C. Lateral

    D. Medial

41. The patella articulates mainly with this bone

A. Femur

B. Tibia

C. Fibula

D. A and B

42. The heart shape of the pelvic inlet occurs mostly in

A. Males

B. Females

C. Children

D. Dwarfs

43. The semitendinosus muscle flexes the knee and can be found on this side of the femur

A. Anterior

B. Posterior

C. Anterior lateral

D. Posterior medial

44. The biceps femoris flexes the knee and can be found on this side of the femur

A. Anterior

B. Posterior

C. Anterior medial

D. Posterior lateral

45. An ankle injury will most likely be result of twisting the foot with this motion

A. Protraction

B. Subtraction

C. Eversion

D. Inversion

46. The vein that is located medially to the thigh and drains the medial thigh, leg, and foot is

A. Femoral

B. Saphenous

C. Popliteal

D. Deep femoral

47. The main artery of the femur is the femoral artery and it can be found here on the thigh

A. Anterior

B. Posterior

C. Anterior medial

D. Posterior lateral

48. The artery which supplies the calf, peroneal muscles, and the ankle with blood would be called

A. Femoral

B. Popliteal

C. Fibular

D. Tibial

49. The artery which supplies the knee, leg, foot, and the digits of the foot with blood would be called

A. Femoral

B. Popliteal

C. Fibular

D. Tibial

50. Which major arteries supply the lower limbs with blood?

A. Internal iliacs

B. External iliacs

C. Femoral major

D. Femoral minor

51. A ROM test is generally used on all joint replacement procedures

A. True

B. False

52. The most common indication for a hip replacement is Degenerative arthritis

A. True

B. False

53. When dislocating a hip joint one person should pry on the femur

A. True

B. False

54. Anterior lateral approach hip surgery will need an abduction pillow

A. True

B. False

55. During a hip replacement all the bones involved are in danger of fracture

A. True

B. False

56. During a hip replacement all these nerves are in danger of damage sciatic, femoral, and peroneal

A. True

B. False

57. The largest bone in the body is the skull

A. True

B. False

58. The largest nerve in the body is the femoral

A. True

B. False

59. The lumbar plexus includes L1-L4 and the sacral plexus encompasses L4-S3

A. True

B. False

60. The femoral or sciatic nerve can be damaged during surgery by the position of an extremity, instrument trauma, or surgical trauma

A. True

B. False

61. If the sciatic nerve is damaged during surgery it will take at least a year to recover from the trauma

A. True

B. False

62. The sciatic nerve is at risk when the assistant is retracting the iliopsoas muscle and tendon

A. True

B. False

63. The sciatic nerve is at risk during a routine hip posterior capsulotomy

A. True

B. False

64. The peroneal nerve is the nerve most likely to be injured when dropping the foot down during a hip procedure

A. True

B. False

65. The sciatic nerve is easily identified by the fatty surrounding tissue

A. True

B. False

66. When the assistant takes measures to avoid injury to the femoral nerve they will also be protecting the deep femoral artery and vein

A. True

B. False

67. When retracting in the area of a nerve, vein, or artery one should use a blunt retractor

A. True

B. False

68. Removal of the femoral head without replacement is termed arthroplasty

A. True

B. False

69. Older age, obesity, and diabetes are increased risk factors for infection in an arthroplasty

A. True

B. False

70. The fact that a patient has a urinary infection should cancel an operation that includes implants

A. True

B. False

71. Antibiotics used in cement will leach out locally for about a year

    A. True

    B. False

72. Organisms that infect joint replacements can be either gram positive or gram negative

    A. True

    B. False

73. When one uses a Charley incision the hip will be dislocated towards the anterior aspect

    A. True

    B. False

74. Arthritis, dislocation, or girdlestone are disorders that would warrant a hip replacement

    A. True

    B. False

75. For a younger person who is getting a hip replacemen, a metal on metal implant is recommended

    A. True

    B. False

76. The weakest link in a hip procedure is cement

    A. True

    B. False

77. The actual load that a metal can withstand without bending or breaking is called fatigue limit

A. True

B. False

78. The safest tourniquet to use for an extremity is the pneumatic

A. True

B. False

79. An extremity should be held up for at least two minuets to express the blood

A. True

B. False

80. An adult arm tourniquet should be no more than 500 mm Hg

A. True

B. False

81. An adult thigh tourniquet should be no more than 500 mm Hg

A. True

B. False

82. A safe time limit on an arm tourniquet normally should be one hour

A. True

B. False

83. A safe time limit on a thigh tourniquet normally should be one and a half hours

A. True

B. False

84. The fibula articulates mainly with the tibia

A. True

B. False

85. The capitate bone is not in the foot

A. True

B. False

86. The turning of the foot outward is termed protraction

A. True

B. False

87. The turning of the foot inward is termed inversion

A. True

B. False

88. The greater sciatic notch is located on the posterior surface of the ilium

A. True

B. False

89. The obturator foramen is located on the medial surface of the ilium

A. True

B. False

90. The femoral epicondyles are located on the lateral condyle only

A. True

B. False

91. The patella articulates mainly with the femur bone

A. True

B. False

92. The heart shape of the pelvic inlet occurs mostly in males

A. True

B. False

93. The semitendinosus muscle flexes the knee and can be found on the posterior medial side of the femur

A. True

B. False

94. The biceps femoris flexes the knee and can be found on the posterior lateral side of the femur

A. True

B. False

95. An ankle injury will most likely be result of twisting the foot with the inversion motion

A. True

B. False

96. The vein that is located medially to the thigh and drains medial thigh, leg, and foot is the deep femoral

    A. True

    B. False

97. The main artery of the femur is the femoral artery and it can be found anterior medial on the thigh

    A. True

    B. False

98. The artery which supplies the calf, peroneal muscles, and the ankle with blood would be the fibular

    A. True

    B. False

99. The artery which supplies the knee, leg, foot, and the digits of the foot with blood would be the fibular artery

    A. True

    B. False

100. The external iliacs are the major arteries that supply the lower limbs with blood

    A. True

    B. False

1 A ❑ B ❑ C ❑ D ❑
2 A ❑ B ❑ C ❑ D ❑
3 A ❑ B ❑ C ❑ D ❑
4 A ❑ B ❑ C ❑ D ❑
5 A ❑ B ❑ C ❑ D ❑
6 A ❑ B ❑ C ❑ D ❑
7 A ❑ B ❑ C ❑ D ❑
8 A ❑ B ❑ C ❑ D ❑
9 A ❑ B ❑ C ❑ D ❑
10 A ❑ B ❑ C ❑ D ❑
11 A ❑ B ❑ C ❑ D ❑
12 A ❑ B ❑ C ❑ D ❑
13 A ❑ B ❑ C ❑ D ❑
14 A ❑ B ❑ C ❑ D ❑
15 A ❑ B ❑ C ❑ D ❑
16 A ❑ B ❑ C ❑ D ❑
17 A ❑ B ❑ C ❑ D ❑
18 A ❑ B ❑ C ❑ D ❑
19 A ❑ B ❑ C ❑ D ❑
20 A ❑ B ❑ C ❑ D ❑
21 A ❑ B ❑ C ❑ D ❑
22 A ❑ B ❑ C ❑ D ❑
23 A ❑ B ❑ C ❑ D ❑
24 A ❑ B ❑ C ❑ D ❑
25 A ❑ B ❑ C ❑ D ❑
26 A ❑ B ❑ C ❑ D ❑
27 A ❑ B ❑ C ❑ D ❑
28 A ❑ B ❑ C ❑ D ❑
29 A ❑ B ❑ C ❑ D ❑
30 A ❑ B ❑ C ❑ D ❑
31 A ❑ B ❑ C ❑ D ❑
32 A ❑ B ❑ C ❑ D ❑
33 A ❑ B ❑ C ❑ D ❑
34 A ❑ B ❑ C ❑ D ❑
35 A ❑ B ❑ C ❑ D ❑
36 A ❑ B ❑ C ❑ D ❑
37 A ❑ B ❑ C ❑ D ❑
38 A ❑ B ❑ C ❑ D ❑
39 A ❑ B ❑ C ❑ D ❑
40 A ❑ B ❑ C ❑ D ❑
41 A ❑ B ❑ C ❑ D ❑
42 A ❑ B ❑ C ❑ D ❑
43 A ❑ B ❑ C ❑ D ❑
44 A ❑ B ❑ C ❑ D ❑
45 A ❑ B ❑ C ❑ D ❑
46 A ❑ B ❑ C ❑ D ❑
47 A ❑ B ❑ C ❑ D ❑
48 A ❑ B ❑ C ❑ D ❑
49 A ❑ B ❑ C ❑ D ❑
50 A ❑ B ❑ C ❑ D ❑
51 A ❑ B ❑ C ❑ D ❑
52 A ❑ B ❑ C ❑ D ❑
53 A ❑ B ❑ C ❑ D ❑
54 A ❑ B ❑ C ❑ D ❑
55 A ❑ B ❑ C ❑ D ❑
56 A ❑ B ❑ C ❑ D ❑
57 A ❑ B ❑ C ❑ D ❑
58 A ❑ B ❑ C ❑ D ❑
59 A ❑ B ❑ C ❑ D ❑
60 A ❑ B ❑ C ❑ D ❑
61 A ❑ B ❑ C ❑ D ❑
62 A ❑ B ❑ C ❑ D ❑
63 A ❑ B ❑ C ❑ D ❑
64 A ❑ B ❑ C ❑ D ❑
65 A ❑ B ❑ C ❑ D ❑
66 A ❑ B ❑ C ❑ D ❑
67 A ❑ B ❑ C ❑ D ❑
68 A ❑ B ❑ C ❑ D ❑
69 A ❑ B ❑ C ❑ D ❑
70 A ❑ B ❑ C ❑ D ❑
71 A ❑ B ❑ C ❑ D ❑
72 A ❑ B ❑ C ❑ D ❑
73 A ❑ B ❑ C ❑ D ❑
74 A ❑ B ❑ C ❑ D ❑
75 A ❑ B ❑ C ❑ D ❑
76 A ❑ B ❑ C ❑ D ❑
77 A ❑ B ❑ C ❑ D ❑
78 A ❑ B ❑ C ❑ D ❑
79 A ❑ B ❑ C ❑ D ❑
80 A ❑ B ❑ C ❑ D ❑
81 A ❑ B ❑ C ❑ D ❑
82 A ❑ B ❑ C ❑ D ❑
83 A ❑ B ❑ C ❑ D ❑
84 A ❑ B ❑ C ❑ D ❑
85 A ❑ B ❑ C ❑ D ❑
86 A ❑ B ❑ C ❑ D ❑
87 A ❑ B ❑ C ❑ D ❑
88 A ❑ B ❑ C ❑ D ❑
89 A ❑ B ❑ C ❑ D ❑
90 A ❑ B ❑ C ❑ D ❑
91 A ❑ B ❑ C ❑ D ❑
92 A ❑ B ❑ C ❑ D ❑
93 A ❑ B ❑ C ❑ D ❑
94 A ❑ B ❑ C ❑ D ❑
95 A ❑ B ❑ C ❑ D ❑
96 A ❑ B ❑ C ❑ D ❑
97 A ❑ B ❑ C ❑ D ❑
98 A ❑ B ❑ C ❑ D ❑
99 A ❑ B ❑ C ❑ D ❑
100 A ❑ B ❑ C ❑ D ❑

CH7 Hip Revision

Multiple Choice

1. D	11. B	21. C	31. B	41. A
2. C	12. B	22. D	32. B	42. A
3. A	13. A	23. B	33. B	43. D
4. B	14. C	24. D	34. B	44. D
5. D	15. B	25. C	35. D	45. D
6. D	16. B	26. B	36. C	46. B
7. A	17. B	27. A	37. D	47. C
8. B	18. D	28. C	38. B	48. C
9. C	19. D	29. B	39. D	49. D
10. D	20. C	30. A	40. B	50. B

True / False

51. True	61. True	71. Three weeks	81. True	91. True
52. True	62. Femoral	72. True	82. True	92. True
53. True	63 True	73. Posterior	83. True	93. True
54. Posterior lateral	64. True	74. True	84. True	94. True
55. True	65. True	75. Ceramic	85. True	95. True
56. True	66. True	76. True	86. Eversion	96. Saphenous
57. Femur	67. True	77. True	87. True	97. True
58. Sciatic	68. Girdlestone	78. True	88. True	98. True
59. True	69. True	79. True	89. True	99. Tibial
60. True	70. True	80. 300	90. Both condyles	100. True

CHAPTER 8

THYROIDECTOMY CASE STUDY

Thyroidectomy

This afternoon the first case on the board for us is a thyroidectomy. I will need to stop by x-ray and pick up the patients folder containing x-rays and cat scans. On my way back I will check the chart for any known allergies, previous surgeries, or any other abnormalities. I will also speak to the patient, identify myself and ask if there are any questions that she might have regarding the surgery. I will make the team aware of anything extra that they may need. Once I get to the room, I will then check that the room is ready for the patient: x-rays hung, proper drugs available, the case is open and all instrumentation is on hand and working properly, that anesthesia has spoken to the patient and we are ready to bring the patient to the room. The reason we are doing this surgery today is that the patient has hyperthyroidism. This is causing pain and restricting air flow.

While the anesthesiologist is sedating the patient, I will assist with intubations by applying cricoid pressure (Selleck's procedure) if needed. With the patient asleep I can now place the Foley, and since this is a female I will only need to insert the catheter about four inches. Now we prepare to place the patient into modified supine position. With the help of all team members, we will slowly raise our patient's head and place a roll under the shoulders with the arms tucked. While anesthesia protects the head and airway, we will place the head in a headrest and tape it down securely so it will not move during the surgery.

Now I will need to place the bovie pad over a fatty location but still close to incision site. I will carefully place warm blankets over the arms and legs of my patient to prevent hypothermia. The table strap is placed last to prevent the patient from a fall when the table is repositioned. With the site properly marked I can beep the surgeon and go scrub while the circulator does the prep. The Surgical technologist and I will drape the patient and hook up bovie and suction while the surgeon is scrubbing. A surgical timeout is done to identify the proper patient with their surgery and we can begin.

The incision we are going to use is going to be symmetrical and transverse across the Langer lines. We only extend the incision as to get the proper exposer. I will use skin rakes, as the superficial layers of tissue need to be exposed.

With use of the bovie, the fascia is freed from the platysma muscle and the platysma is divided with blunt and sharp dissection. I will then place a Weitlaner retractor into the wound to expose the underlying muscles and thyroid. The strap muscles will then be divided and retracted off of the thyroid. As we expose the middle thyroid vein it will need to be clamped, cut and ligated. Now as we dissect the superior poles we can locate, divide, and ligate the superior vessels. While all of our blunt and sharp dissection is being done we must be constant in our identifying of the parathyroid glands and the recurrent laryngeal nerve. The same procedure can now be done with the inferior vessels. The thyroid is dissected free from the trachea and now can be clamped, cut, ligated, and freed along with the isthmus and pyramidal lobe.

We can check for bleeding and verify that the recurrent nerve has not been damaged by this procedure. Last we are going to irrigate with saline, place our drain and prepare to close. For my closure I will use a 3-0 Vicryl for strength and interrupt those, next a 4-0 Monocryl to finish the closure sub cue. A 3-0 nylon is used for the drain and after an injection of 10cc of Marcaine I will apply steri strips and mastasol, finishing with antibiotic ointment, adptic, fluffs, and a Queen Anne's dressing.

The patient can now be extubated. Care will be taken as we proceed to move the patient to her stretcher and transported to recovery. On my follow up I will advise the patient on dressing changes, showers, and any neck mobilization. The expected out come will be to return to normal activities after six to eight weeks.

Case study

In this section you will need to describe a Thyroidectomy in detail from your perspective (one that you have done in the past). List any differences in positioning, drugs used, or incision, etc.

1. The thyroid gland secretes which hormone?
   A. Thyroxine
   B. Lipotropin
   C. Thyroid-stimulating
   D. Prolactin

2. The parathyroid glands secretes which hormone?
   A. Thyroxine
   B. Parathyroid-hormone
   C. Calcitonin
   D. Prolactin

3. Calcium levels are maintained by this gland
   A. Hypothalamus
   B. Thyroid
   C. Pituitary
   D. Parathyroid

4. The hypothalamus secretes this hormone
   A. Thyroxine
   B. Calcitonin
   C. Thyroid-releasing
   D. Prolactin

5. Growth hormone is secreted by this gland

   A. Thyroid

   B. Parathyroid

   C. Pituitary

   D. Hypothalamus

6. The thymus in the chest secrets this hormone

   A. Thymosin

   B. Melatonin

   C. Thyroxine

   D. Prolactin

7. Epinephrine is produced in this gland

   A. Liver

   B. Thymus

   C. Adrenal

   D. Pancreas

8. The testes will produce this hormone

   A. Inhibin

   B. Testosterone

   C. Relaxin

   D. A and B

9. The recurrent laryngeal nerve branches off of this main nerve

    A. Cervical

    B. Vagus

    C. Sural

    D. Thoracic

10. The heart and lungs are innervated by this nerve

    A. Cervical

    B. Vagus

    C. Sural

    D. Thoracic

11. There is normally how many parathyroid glands present in the human body?

    A. Two

    B. Four

    C. Six

    D. Four pair

12. Parathyroid glands are usually found on this plane of the thyroid

    A. Anterior

    B. Posterior

    C. Medially

    D. Caudally

13. In relation to the thyroid, the trachea is located

A. Anterior

B. Posterior

C. Lateral

D. Medial

14. The tissue that connects the two halves of the thyroid are called

A. Infundibulum

B. Isthmus

C. Ampulla

D. Pyramidal

15. On the thyroid, there is a tissue process that occurs in about half of all cases. It is called

A. Infundibulum

B. Isthmus

C. Ampulla

D. Pyramidal

16. The name of the bone in the neck that does not articulate with another bone is the

A. Cornua

B. Malleus

C. Hyoid

D. Stylohyoid

17. The lower jaw bone is called the

    A. Maxillae

    B. Palatine

    C. Mandible

    D. Zygomatic

18. The upper jaw bone is called the

    A. Maxillae

    B. Palatine

    C. Mandible

    D. Zygomatic

19. These two bones are considered to be the cheek bones

    A. Maxillae

    B. Palatine

    C. Mandible

    D. Zygomatic

20. The Adam's apple is formed from this structure

    A. Stylohyoid

    B. Cornu

    C. Hyoid

    D. Thyroid cartilage

21. Paralysis of the vocal cords and dysphagia can be traced to the damage of this nerve

A. Glossopharyngeal
B. Vagus
C. Accessory
D. Hypoglossal

22. Hyper activity of the thyroid gland can result in this disorder

A. Addison's
B. Hypoglycemia
C. Diabetes
D. Grave's disease

23. Hypoparathyroidism will cause this disorder

A. Calcium imbalance
B. Hypoglycemia
C. Diabetes
D. Myxedema

24. This disease is caused by the decreased production of insulin in the pancreas

A. Addison's
B. Hypoglycemia
C. Diabetes
D. Grave's disease

25. This muscle lies directly on the thyroid and will be retracted during a thyroidectomy

A. Omohyoid

B. Sternohyoid

C. Cricothyroid

D. Mylohyoid

26. These muscles of the neck are responsible for the major movements of the head

A. Sternocleidomastoid

B. Infraspinatus

C. Deltoid

D. Platysma

27. Due to the vascular nature of the neck this local anesthesia should be injected in to the skin

A. Lidocaine

B. Lidocaine with epi

C. Marcaine

D. Procaine

28. With the possibility of devascularizing tissue such as a nose, this drug should not be injected in to the skin

A. Lidocaine

B. Lidocaine with epi

C. Marcaine

D. Procaine

29. The Latin term non per os is the same as

A. Nothing by mouth

B. Nil per os

C. NPO

D. All the above

30. Due to the tissue structures you will encounter during a thyroid-ectomy, you should use this kind of needle

A. Straight

B. Tapered

C. Cutting

D. Reverse cutting

31. On a thyroidectomy incision this would be the closure material of choice

A. Silk

B. Cotton

C. Nylon

D. Staples

32. The lines that follow the natural contour of the body are called

A. Kelly's lines

B. Langer's lines

C. Rossi's lines

D. Sanford's lines

33. The bilobed lymphatic gland located next to the heart is the

A. Lacteal

B. Spleen

C. Thymus

D. Thyroid

34. The organ located in the superior region of the abdomen needed to filter blood is the

A. Lacteal

B. Spleen

C. Thymus

D. Thyroid

35. The glands found in the small intestines that filter fat from the digestive tract are

A. Lacteal

B. Spleen

C. Thymus

D. Thyroid

36. The Bubonic plague that ran rampant in medieval times was bacteria that localized in this tissue

A. Brain

B. Heart

C. Skin

D. Lymph

37. For this reason, a tracheostomy would be indicated

    A. Throat trauma

    B. Irradiation of the neck

    C. Air hunger

    D. Any of the above

38. Repair of a complete fracturing of the maxilla, zygomatic arch, and nasal bones would be considered

    A. Le Fort I

    B. L Fort II

    C. Le Fort III

    D. Caldwell Luc

39. The surgery performed for infection of the maxillary sinus is

    A. Le Fort I

    B. L Fort II

    C. Le Fort III

    D. Caldwell Luc

40. The nerve that we use meticulous dissection in finding during a Parotidectomy is the

    A. Facial

    B. Laryngeal

    C. Vagus

    D. All of the above

41. For a radical neck dissection we use this incision

    A. MacFee

    B. Crile

    C. Martin

    D. Any one of the above

42. A patient with difficulty in swallowing would be considered to have

    A. Varices

    B. Dysphagia

    C. Apnea

    D. GERD

43. A patient with difficulty in breathing would be considered to have

    A. Varices

    B. Dysphagia

    C. Apnea

    D. GERD

44. The v-shaped dressing used for a thyroid procedure should be

    A. Queen Anne's

    B. Pressure

    C. Wet to dry

    D. V-dressing

45. The dressing used for an incision that is going to be left open to heal is

A. Queen Anne's

B. Pressure

C. Wet to dry

D. V-dressing

46. The deadly condition caused by removing all the parathyroid glands is called

A. Fibroma

B. Tetany

C. Lacus

D. Granuloma

47. The term "Above all, do no harm." translates from

A. Res ipsa loquitu

B. Respondeat superior

C. Primum non nocere

D. Aeger primo

48. The term "Let the master answer;" translates from

A. Res ipsa loquitu

B. Respondeat superior

C. Primum non nocere

D. Aeger primo

49. The term "The thing speaks for itself" translates from

    A. Res ipsa loquitu

    B. Respondeat superior

    C. Primum non nocere

    D. Aeger primo

50. The term "The patient first" translates from

    A. Res ipsa loquitu

    B. Respondeat superior

    C. Primum non nocere

    D. Aeger primo

51. The thyroid gland secretes thyroxine hormone

    A. True

    B. False

52. The parathyroid glands secretes thyroxine hormone

    A. True

    B. False

53. Calcium levels are maintained by the parathyroid gland

    A. True

    B. False

54. The hypothalamus secretes thyroid-releasing hormone

    A. True

    B. False

55. Growth hormone is secreted by the thyroid gland

    A. True

    B. False

56. The thymus in the chest secrets thymosin hormone

    A. True

    B. False

57. Epinephrine is produced in the adrenal gland

    A. True

    B. False

58. The testes will produce inhibin and testosterone hormone

    A. True

    B. False

59. The recurrent laryngeal nerve branches off of the sural nerve

    A. True

    B. False

60. The heart and lungs are innervated by the vagus nerve

    A. True

    B. False

61. There are normally four parathyroid glands present in the human body

    A. True

    B. False

62. Parathyroid glands are usually found on the posterior plane of the thyroid

A. True

B. False

63. In relation to the thyroid, the trachea is located posterior

A. True

B. False

64. The tissue that connects the two halves of the thyroid is called isthmus

A. True

B. False

65. On the thyroid, there is a tissue process that occurs in about half of all cases. It is called pyramidal

A. True

B. False

66. The name of the bone in the neck that does not articulate with another bone is the malleus

A. True

B. False

67. The lower jaw bone is called mandible

A. True

B. False

68. The upper jaw bone is called mandible

    A. True

    B. False

69. The two zygomatic bones are considered to be the cheek bones

    A. True

    B. False

70. The (Adam's apple) is formed from the thyroid cartilage

    A. True

    B. False

71. Paralysis of the vocal cords and dysphagia can be traced to the damage of the hypoglossal nerve

    A. True

    B. False

72. Hyperactivity of the thyroid gland can result in Grave's disease

    A. True

    B. False

73. Hypoparathyroidism will cause a calcium imbalance

    A. True

    B. False

74. Diabetes disease is caused by the decreased production of insulin in the pancreas

A. True

B. False

75. The sternohyoid muscle lies directly on the thyroid and will be retracted during a thyroidectomy

A. True

B. False

76. The deltoid muscles of the neck are responsible for the major movements of the head

A. True

B. False

77. Due to the vascular nature of the neck, local anesthesia Lidocaine with epi should be injected in to the skin

A. True

B. False

78. With the possibility of devascularizing tissue such as a nose, Lidocaine with epi should not be injected in to the skin

A. True

B. False

79. The Latin term non per os is also NPO

A. True

B. False

80. Due to the tissue structures you will encounter during a thyroid-ectomy you should use a cutting needle

A. True

B. False

81. On a thyroidectomy incision nylon would be the closure material of choice

A. True

B. False

82. The lines that follow the natural contour of the body are called Langer's lines

A. True

B. False

83. The bilobed lymphatic gland located next to the heart is the thyroid

A. True

B. False

84. The organ located in the superior region of the abdomen needed to filter blood is the spleen

A. True

B. False

85. The glands found in the small intestines that filter fat from the digestive tract are the lacteals

A. True

B. False

86. The Bubonic plague that ran rampant in medieval times was caused by bacteria that localized in the lymphatics

A. True

B. False

87. Irradiation is one of the reasons a tracheostomy would be indicated

A. True

B. False

88. Repair of a complete fracturing of the maxilla, zygomatic arch, and nasal bones would be considered a Le Fort III

A. True

B. False

89. The surgery performed for infection of the maxillary sinus is a Le Fort III

A. True

B. False

90. The nerve that we use meticulous dissection in finding during a Parotidectomy is the facial nerve

A. True

B. False

91. For a radical neck dissection a Crile incision is one of many that we would use

A. True

B. False

92. A patient with difficulty in swallowing would be considered to have dysphagia

    A. True

    B. False

93. A patient with difficulty in breathing would be considered to have dysphagia

    A. True

    B. False

94. The v-shaped dressing used for a thyroid procedure should be a Queen Anne's

    A. True

    B. False

95. The dressing used for an incision that is going to be left open to heal is a Queen Anne's

    A. True

    B. False

96. The deadly condition caused by removing all the parathyroid glands is called lacus

    A. True

    B. False

97. The term "Above all, do no harm." is translated from primum non nocere

    A. True

    B. False

98. The term "Let the master answer;" is translated from respondeat superior

A. True

B. False

99. The term "The thing speaks for itself;" is translated from res ipsa loquitu

A. True

B. False

100. The term "The patient first" is translated form aeger primo

A. True

B. False

1 A ❑ B ❑ C ❑ D ❑
2 A ❑ B ❑ C ❑ D ❑
3 A ❑ B ❑ C ❑ D ❑
4 A ❑ B ❑ C ❑ D ❑
5 A ❑ B ❑ C ❑ D ❑
6 A ❑ B ❑ C ❑ D ❑
7 A ❑ B ❑ C ❑ D ❑
8 A ❑ B ❑ C ❑ D ❑
9 A ❑ B ❑ C ❑ D ❑
10 A ❑ B ❑ C ❑ D ❑
11 A ❑ B ❑ C ❑ D ❑
12 A ❑ B ❑ C ❑ D ❑
13 A ❑ B ❑ C ❑ D ❑
14 A ❑ B ❑ C ❑ D ❑
15 A ❑ B ❑ C ❑ D ❑
16 A ❑ B ❑ C ❑ D ❑
17 A ❑ B ❑ C ❑ D ❑
18 A ❑ B ❑ C ❑ D ❑
19 A ❑ B ❑ C ❑ D ❑
20 A ❑ B ❑ C ❑ D ❑
21 A ❑ B ❑ C ❑ D ❑
22 A ❑ B ❑ C ❑ D ❑
23 A ❑ B ❑ C ❑ D ❑
24 A ❑ B ❑ C ❑ D ❑
25 A ❑ B ❑ C ❑ D ❑
26 A ❑ B ❑ C ❑ D ❑
27 A ❑ B ❑ C ❑ D ❑
28 A ❑ B ❑ C ❑ D ❑
29 A ❑ B ❑ C ❑ D ❑
30 A ❑ B ❑ C ❑ D ❑
31 A ❑ B ❑ C ❑ D ❑
32 A ❑ B ❑ C ❑ D ❑
33 A ❑ B ❑ C ❑ D ❑
34 A ❑ B ❑ C ❑ D ❑
35 A ❑ B ❑ C ❑ D ❑
36 A ❑ B ❑ C ❑ D ❑
37 A ❑ B ❑ C ❑ D ❑
38 A ❑ B ❑ C ❑ D ❑
39 A ❑ B ❑ C ❑ D ❑
40 A ❑ B ❑ C ❑ D ❑
41 A ❑ B ❑ C ❑ D ❑
42 A ❑ B ❑ C ❑ D ❑
43 A ❑ B ❑ C ❑ D ❑
44 A ❑ B ❑ C ❑ D ❑
45 A ❑ B ❑ C ❑ D ❑
46 A ❑ B ❑ C ❑ D ❑
47 A ❑ B ❑ C ❑ D ❑
48 A ❑ B ❑ C ❑ D ❑
49 A ❑ B ❑ C ❑ D ❑
50 A ❑ B ❑ C ❑ D ❑
51 A ❑ B ❑ C ❑ D ❑
52 A ❑ B ❑ C ❑ D ❑
53 A ❑ B ❑ C ❑ D ❑
54 A ❑ B ❑ C ❑ D ❑
55 A ❑ B ❑ C ❑ D ❑
56 A ❑ B ❑ C ❑ D ❑
57 A ❑ B ❑ C ❑ D ❑
58 A ❑ B ❑ C ❑ D ❑
59 A ❑ B ❑ C ❑ D ❑
60 A ❑ B ❑ C ❑ D ❑
61 A ❑ B ❑ C ❑ D ❑
62 A ❑ B ❑ C ❑ D ❑
63 A ❑ B ❑ C ❑ D ❑
64 A ❑ B ❑ C ❑ D ❑
65 A ❑ B ❑ C ❑ D ❑
66 A ❑ B ❑ C ❑ D ❑
67 A ❑ B ❑ C ❑ D ❑
68 A ❑ B ❑ C ❑ D ❑
69 A ❑ B ❑ C ❑ D ❑
70 A ❑ B ❑ C ❑ D ❑
71 A ❑ B ❑ C ❑ D ❑
72 A ❑ B ❑ C ❑ D ❑
73 A ❑ B ❑ C ❑ D ❑
74 A ❑ B ❑ C ❑ D ❑
75 A ❑ B ❑ C ❑ D ❑
76 A ❑ B ❑ C ❑ D ❑
77 A ❑ B ❑ C ❑ D ❑
78 A ❑ B ❑ C ❑ D ❑
79 A ❑ B ❑ C ❑ D ❑
80 A ❑ B ❑ C ❑ D ❑
81 A ❑ B ❑ C ❑ D ❑
82 A ❑ B ❑ C ❑ D ❑
83 A ❑ B ❑ C ❑ D ❑
84 A ❑ B ❑ C ❑ D ❑
85 A ❑ B ❑ C ❑ D ❑
86 A ❑ B ❑ C ❑ D ❑
87 A ❑ B ❑ C ❑ D ❑
88 A ❑ B ❑ C ❑ D ❑
89 A ❑ B ❑ C ❑ D ❑
90 A ❑ B ❑ C ❑ D ❑
91 A ❑ B ❑ C ❑ D ❑
92 A ❑ B ❑ C ❑ D ❑
93 A ❑ B ❑ C ❑ D ❑
94 A ❑ B ❑ C ❑ D ❑
95 A ❑ B ❑ C ❑ D ❑
96 A ❑ B ❑ C ❑ D ❑
97 A ❑ B ❑ C ❑ D ❑
98 A ❑ B ❑ C ❑ D ❑
99 A ❑ B ❑ C ❑ D ❑
100 A ❑ B ❑ C ❑ D ❑

CH8 Thyroidectomy

Multiple Choice

1. A	11. B	21. B	31. C	41. D
2. B	12. B	22. D	32. C	42. B
3. D	13. B	23. A	33. C	43. C
4. C	14. B	24. C	34. B	44. A
5. C	15. D	25. B	35. A	45. C
6. A	16. C	26. A	36. D	46. B
7. D	17. C	27. B	37. D	47. C
8. D	18. A	28. B	38. C	48. B
9. B	19. D	29. D	39. D	49. A
10. B	20. D	30. B	40. A	50. D

True / False

51. True	61. True	71. Vagus	81. True	91. True
52. Calcitonin	62. True	72. True	82. True	92. True
53. True	63. True	73. True	83. Thymus	93. Apnea
54. True	64. True	74. True	84. True	94. True
55. Pituitary	65. True	75. True	85. True	95. Wet to dry
56. True	66. Hyoid	76. Sternocleidomastoid	86. True	96. Tetany
57. True	67. True	77. True	87. True	97. True
58. True	68. Maxilla	78. True	88. True	98. True
59. Vagus	69. True	79. True	89. Caldwell Luc	99. True
60. True	70. True	80. Taper	90. True	100. True

CHAPTER 9

THORACOTOMY CASE STUDY

Thoracotomy

This morning the first case on the board for us is a thoracotomy for a neoplastic growth. I will need to stop by x-ray and pick up the patients folder containing x-rays, cat scans, MRIs and ECG's. On my way back, I will check the chart for any known allergies, previous surgeries, or any other abnormalities. I will also speak to the patient, identify myself and ask if there are any questions that she might have regarding the surgery. I will make the team aware of anything extra that they may need. Once I get to the room, I will check that the room is ready for the patient: chest x-rays hung, proper drugs available, the case is open and all instrumentation is on hand and working properly, that anesthesia has spoken to the patient and we are ready to bring the patient to the room.

While the anesthesiologist is sedating the patient I will assist with intubations by applying cricoid pressure if needed. With the patient asleep I can now place the Foley, tape the eyes, and prepare the pads for the elbows as we prepare to flip the patient into lateral position. With the help of all the team members we will slowly move the patient to the edge and roll the patient over, onto her side while anesthesia protects the head and airway. Once over I will place the arm on an extended arm holder and place an axillary roll under the patient's axillary. The head will be placed on a headrest and padding will then be placed under the knees, ankles, and feet with a pillow placed between the patient's knees with one leg flexed and the upper one straight. Once the patient is fixed properly in the lateral frame, I will arrange the Foley to hang at the head of the table.

Now I will need to place the bovie pad over a fatty location but still close to incision site. I will carefully place warm blankets over the arms and legs of my patient to prevent any hypothermia. The table strap is placed last to prevent the patient from a fall when the table is repositioned. With the site properly marked I can beep the surgeon and go scrub while the circulator does the prep. A surgical timeout is done as to recognize the proper patient with their surgery and we can begin.

We will make our incision posterior laterally about 20cm long, between the fifth and sixth rib and place a small rib spreader (self-retaining retractors). With the scalpel along with the bovie we will dissect thru the pleura and down to the distal lobe of the lung.

While the surgeon is doing this, I will use the Allison lung retractor to gain better exposure. Now that we have maximum exposure we place a stapler marginally around the tumor and fire it twice to get our wedge resection free. We will be sure that the laps are removed at the end of the case when the lap count is initialized. Last we are going to irrigate with warm saline and check for air leaks. The anesthetist will perform a Valsalva's maneuver. Next we can place the chest tube and prepare to close.

The surgeon will leave now and leave me to suture the drains and close the wound. For my closure I will use interrupted number 2 vicryl's for the pleural cavity and 0 vicryl on the next facial layers, then switch to 2-0 vicryl for the final dermal layer and staples for the skin. I will use 3-0 nylon for the drain and now we are ready for the dressing which will consist of Neosporin, Adaptec, Kerlex fluffs, ABD pads, and Medipore tape. We will connect the drain tubes and tape those down to the patient. Now we should be ready to carefully move the patient back on to her stretcher for extubation. Depending on the reason for the thoracotomy the outcome should be favorable with the patient able to return to normal activity in four to six weeks.

Case study

In this section you will need to describe a thoracotomy in detail from your perspective (one that you have done in the past). List any differences in positioning, drugs used, or incision, etc.

1. An increase of intrathoracic pressure by forcible exhalation is considered
   A. Quincke's pulse
   B. Valsalva's maneuver
   C. Koch's maneuver
   D. Corrigan's pulse

2. The universal blood donor for all types of blood is
   A. A-
   B. B-
   C. AB+
   D. 0-

3. The percentage of whole blood that is made up of plasma
   A. 10%
   B. 20%
   C. 55%
   D. 91%

4. The percentage of whole blood plasma that contains water is
   A. 10%
   B. 20%
   C. 55%
   D. 91%

5. The test to determine ABO and Rh of blood groups is termed

   A. Cross match

   B. Blood typing

   C. Blood count

   D. Prothrombin time

6. The test used with blood typing to prevent reactions is

   A. Cross match

   B. CBC

   C. Blood count

   D. Prothrombin time

7. Which blood group is responsible for hemolytic disease of the newborn

   A. AB0

   B. 0+

   C. Rh

   D. None of the above

8. The component of blood responsible for the red color

   A. Hemoglobin

   B. Heme

   C. Oxygen

   D. Carbon dioxide

9. What percentage of red blood cells are replaced every day?
   A. 1%
   B. 10%
   C. 50%
   D. .001%

10. Deoxygenated blood flows to the lungs via this route
    A. Pulmonary arteries
    B. Pulmonary veins
    C. Azygos vein
    D. None of the above

11. Oxygenated blood flows from the lungs via this route
    A. Pulmonary arteries
    B. Pulmonary veins
    C. Azygos vein
    D. None of the above

12. This valve separates the left atrium from the left ventricle in the heart
    A. Tricuspid
    B. Bicuspid
    C. Pulmonary
    D. Aortic

13. This valve separates the right atrium from the right ventricle in the heart

    A. Tricuspid

    B. Bicuspid

    C. Pulmonary

    D. Aortic

14. This node is considered the pacemaker of the heart

    A. Purkinje

    B. Atrioventricular

    C. Sinoatrial

    D. AV

15. In reference to the heart this is the term for the state of the heart at rest

    A. Systole

    B. Diastole

    C. Isovolumic

    D. Fibrillation

16. In reference to the heart this is the term for the state of the heart during contraction

    A. Systole

    B. Diastole

    C. Isovolumic

    D. Fibrillation

17. Which of these components will play an important part in the clotting of blood

    A. Fibrin

    B. Platelets

    C. Thrombin

    D. All the above

18. The condition of or absence of the tricuspid valve is referred to

    A. Transposition

    B. Atresia

    C. Arteriosus

    D. Coarctation

19. The condition of having a single vessel arising from both ventricles is

    A. Transposition

    B. Atresia

    C. Arteriosus

    D. Coarctation

20. Patent ductus arteriosus is a failure of the great vessels in the heart to do what?

    A. Open

    B. Close

    C. Separate

    D. Oxygenate

21. When the surgeon cannulates the heart which vessel is done first?

   A. Aorta

   B. Vena Cava

   C. Coronary

   D. Pulmonary

22. When a blood pump oxygenator is used during a heart procedure from which vessel is the blood taken from

   A. Aorta

   B. Vena Cava

   C. Coronary

   D. Pulmonary

23. When a blood pump oxygenator is used during a heart procedure from which vessel is the blood returned to system in

   A. Aorta

   B. Vena Cava

   C. Coronary

   D. Pulmonary

24. During a heart procedure, this technique is often utilized to reduce oxygen demand

   A. Occlusion

   B. Cannulation

   C. Hyperthermia

   D. Hypothermia

25.In a coronary artery bypass which artery is most commonly used as a replacement?

A. Azygos

B. Mammary

C. Saphenous

D. B and C

26. During a heart procedure this drug is used to prevent vasospasm

A. Benzodiazepines

B. Lidocaine

C. Papaverine

D. Thrombin

27. Life threatening dysrhythmias can be arrested by the IV administration of this drug

A. Benzodiazepines

B. Lidocaine

C. Papaverine

D. Thrombin

28. This emergency drug is given to increase blood pressure

A. Codeine

B. Epinephrine

C. Diazepam

D. Atropine

29. On the ECG and cardiac cycle which part of the wave represents ventricular relaxation?

A. P

B. QRS

C. T

D. U

30. On the ECG and cardiac cycle, which part of the wave represents atrial contraction?

A. P

B. QRS

C. T

D. U

31. This test was designed to measure end-tidal CO_2

A. Capnography

B. Spirometry

C. Oximetry

D. ECG

32. The least traumatic wound closure material used today is

A. Silk

B. Nylon

C. Steel

D. Monocryl

33. For suture that will retain the longest tensile strength you would use
    A. Plain gut
    B. Chromic gut
    C. PDS
    D. Monocryl

34. The suture of choice for an aorta incision closure would be
    A. PDS
    B. Ethibond
    C. Prolene
    D. Chromic gut

35. When approaching the right distal lung which ribs are spread open for access?
    A. Two and three
    B. Five and six
    C. Eleven and twelve
    D. None of these

36. The incision for a thoracotomy is called
    A. Pfannenstiel
    B. Kocker
    C. Mcburney
    D. Horizontal

37. The congenital deformity of the chest that results in a funnel shaped depression is

A. Pleurisies

B. Pectus carinatum

C. Pectus excavatum

D. Thoracic syndrome

38. The congenital deformity of the chest that results in an upward protrusion of the ribs is

A. Pleurisies

B. Pectus carinatum

C. Pectus excavatum

D. Thoracic syndrome

39. This vein is one three major vessels that drain the thoracic cavity

A. Subclavian

B. Axillary

C. Basilic

D. Azygos

40. This muscle of the thorax is used on inspiration and to depress the floor of the thorax

A. Intercostalis

B. Scalenus

C. Diaphragm

D. Scalene

41. On the rib cage there are how many floating ribs total?

A. Two

B. Four

C. Six

D. Five

42. The diaphragm is activated by this nerve

A. Vagus

B. Phrenic

C. Thoracic

D. Diaphragmatic

43. The procedure for evaluation of the nodal involvement, mediastinal masses, or carcinoma of the lung is

A. Bronchoscopy

B. Mediastinoscopy

C. Thoracoscopy

D. Thoracentesis

44. The lung on the right side of the heart has how many lobes

A. One

B. Two

C. Three

D. Four

45. The lung on the left side of the heart has how many lobes

A. One

B. Two

C. Three

D. Four

46. The separation of lung from its attachments is

A. Pleurolith

B. Pleurodynia

C. Pleurocele

D. Pleurolysis

47. Pain in the intercostal muscles is referred to as

A. Pleurolith

B. Pleurodynia

C. Pleurocele

D. Pleurolysis

48. Hernia of lung tissue is referred to as

A. Pleurolith

B. Pleurodynia

C. Pleurocele

D. Pleurolysis

49. The divisions in the lungs are known as

    A. Fissures

    B. Sulcus

    C. Separations

    D. Gyrus

50. Which organ is considered the largest in the body?

    A. Liver

    B. Brain

    C. Uterus

    D. Lungs

51. An increase of intrathoracic pressure by forcible exhalation is considered Valsalva's maneuver

    A. True

    B. False

52. The universal blood donor for all types of blood is 0-

    A. True

    B. False

53. The percentage of whole blood that is made up of plasma is 10%

    A. True

    B. False

54. The percentage of whole blood plasma that contains water is 10%

 A. True

 B. False

55. The test to determine ABO and RH of blood groups is termed blood typing

 A. True

 B. False

56. The test used with blood typing to prevent reactions is cross matching

 A. True

 B. False

57. The RH blood group is responsible for hemolytic disease of the newborn

 A. True

 B. False

58. Heme is the component of blood responsible for the red color

 A. True

 B. False

59. Ten percent of red blood cells are replaced every day

 A. True

 B. False

60. Deoxygenated blood flows to the lungs via the pulmonary veins

 A. True

 B. False

61. Oxygenated blood flows from the lungs via the pulmonary veins

A. True

B. False

62. The tricuspid valve separates the left atrium from the left ventricle in the heart

A. True

B. False

63. The tricuspid valve separates the right atrium from the right ventricle in the heart

A. True

B. False

64. The Purkinje node is considered the pacemaker of the heart

A. True

B. False

65. In reference to the heart diastole is the term for the state of the heart at rest

A. True

B. False

66. In reference to the heart systole is the term for the state of the heart during contraction

A. True

B. False

67. Fibrin, thrombin, and platelets all are components that play an important part in the clotting of blood

    A. True

    B. False

68. The condition of or absence of the tricuspid valve is referred to as atresia

    A. True

    B. False

69. The condition of having a single vessel arising from both ventricles is called transposition

    A. True

    B. False

70. Patent ductus arteriosus is a failure of the great vessels in the heart to close

    A. True

    B. False

71. When the surgeon cannulates the heart the aorta is done first

    A. True

    B. False

72. When a blood pump oxygenator is used during a heart procedure, the aorta is where the blood is taken from the system

    A. True

    B. False

73. When a blood pump oxygenator is used during a heart procedure, the aorta is where the blood is returned to the system

A. True

B. False

74. During a heart procedure, hypothermia is the technique often utilized to reduce oxygen demand

A. True

B. False

75. In a coronary artery bypass the mammary and saphenous veins are most commonly used as a replacement

A. True

B. False

76. During a heart procedure, Papaverine is the drug used to prevent vasospasm

A. True

B. False

77. Life threatening dysrhythmias can be arrested by the IV administration of Lidocaine

A. True

B. False

78. Epinephrine is an emergency drug that is given to increase blood pressure

A. True

B. False

79. On the ECG and cardiac cycle, the U part of the wave represents ventricular relaxation

A. True

B. False

80. On the ECG and cardiac cycle, the T part of the wave represents atrial contraction

A. True

B. False

81. Capnography is a test designed to measure end-tidal CO_2

A. True

B. False

82. The least traumatic wound closure material used today is steel

A. True

B. False

83. For a suture that will retain the longest tensile strength we would use PDS

A. True

B. False

84. The suture of choice for an aorta incision closure would be PDS

A. True

B. False

85. When approaching the right distal lung the fifth and sixth ribs are spread open for access

A. True

B. False

86. The Kocker incision is the incision of choice for a thoracotomy

A. True

B. False

87. Pectus excavatum is the congenital deformity of the chest that results in a funnel shaped depression

A. True

B. False

88. Pectus carinatum is the congenital deformity of the chest that results in an upward protrusion of the ribs

A. True

B. False

89. The axillary vein is one of three major vessels that drain the thoracic cavity

A. True

B. False

90. The diaphragm muscle of the thorax is used on inspiration and to depress the floor of the thorax

A. True

B. False

91. On the rib cage there are six floating ribs total

A. True

B. False

92. The diaphragm is activated by the vagus nerve

A. True

B. False

93. Mediastinoscopy is the procedure for evaluation of the nodal involvement, mediastinal masses, or carcinoma of the lung

A. True

B. False

94. The lung on the right side of the heart has three lobes

A. True

B. False

95. The lung on the left side of the heart has three lobes

A. True

B. False

96. Pleurolysis is the separation of lung from its attachments is

A. True

B. False

97. Pain in the intercostal muscles is referred to as pleurodynia

    A. True

    B. False

98. Hernia of lung tissue is referred to pleurolith

    A. True

    B. False

99. The divisions in the lungs are known fissures

    A. True

    B. False

100. The lungs are considered the largest organs in the body

    A. True

    B. False

1 A ❑ B ❑ C ❑ D ❑
2 A ❑ B ❑ C ❑ D ❑
3 A ❑ B ❑ C ❑ D ❑
4 A ❑ B ❑ C ❑ D ❑
5 A ❑ B ❑ C ❑ D ❑
6 A ❑ B ❑ C ❑ D ❑
7 A ❑ B ❑ C ❑ D ❑
8 A ❑ B ❑ C ❑ D ❑
9 A ❑ B ❑ C ❑ D ❑
10 A ❑ B ❑ C ❑ D ❑
11 A ❑ B ❑ C ❑ D ❑
12 A ❑ B ❑ C ❑ D ❑
13 A ❑ B ❑ C ❑ D ❑
14 A ❑ B ❑ C ❑ D ❑
15 A ❑ B ❑ C ❑ D ❑
16 A ❑ B ❑ C ❑ D ❑
17 A ❑ B ❑ C ❑ D ❑
18 A ❑ B ❑ C ❑ D ❑
19 A ❑ B ❑ C ❑ D ❑
20 A ❑ B ❑ C ❑ D ❑
21 A ❑ B ❑ C ❑ D ❑
22 A ❑ B ❑ C ❑ D ❑
23 A ❑ B ❑ C ❑ D ❑
24 A ❑ B ❑ C ❑ D ❑
25 A ❑ B ❑ C ❑ D ❑
26 A ❑ B ❑ C ❑ D ❑
27 A ❑ B ❑ C ❑ D ❑
28 A ❑ B ❑ C ❑ D ❑
29 A ❑ B ❑ C ❑ D ❑
30 A ❑ B ❑ C ❑ D ❑
31 A ❑ B ❑ C ❑ D ❑
32 A ❑ B ❑ C ❑ D ❑
33 A ❑ B ❑ C ❑ D ❑
34 A ❑ B ❑ C ❑ D ❑
35 A ❑ B ❑ C ❑ D ❑
36 A ❑ B ❑ C ❑ D ❑
37 A ❑ B ❑ C ❑ D ❑
38 A ❑ B ❑ C ❑ D ❑
39 A ❑ B ❑ C ❑ D ❑
40 A ❑ B ❑ C ❑ D ❑
41 A ❑ B ❑ C ❑ D ❑
42 A ❑ B ❑ C ❑ D ❑
43 A ❑ B ❑ C ❑ D ❑
44 A ❑ B ❑ C ❑ D ❑
45 A ❑ B ❑ C ❑ D ❑
46 A ❑ B ❑ C ❑ D ❑
47 A ❑ B ❑ C ❑ D ❑
48 A ❑ B ❑ C ❑ D ❑
49 A ❑ B ❑ C ❑ D ❑
50 A ❑ B ❑ C ❑ D ❑
51 A ❑ B ❑ C ❑ D ❑
52 A ❑ B ❑ C ❑ D ❑
53 A ❑ B ❑ C ❑ D ❑
54 A ❑ B ❑ C ❑ D ❑
55 A ❑ B ❑ C ❑ D ❑
56 A ❑ B ❑ C ❑ D ❑
57 A ❑ B ❑ C ❑ D ❑
58 A ❑ B ❑ C ❑ D ❑
59 A ❑ B ❑ C ❑ D ❑
60 A ❑ B ❑ C ❑ D ❑
61 A ❑ B ❑ C ❑ D ❑
62 A ❑ B ❑ C ❑ D ❑
63 A ❑ B ❑ C ❑ D ❑
64 A ❑ B ❑ C ❑ D ❑
65 A ❑ B ❑ C ❑ D ❑
66 A ❑ B ❑ C ❑ D ❑
67 A ❑ B ❑ C ❑ D ❑
68 A ❑ B ❑ C ❑ D ❑
69 A ❑ B ❑ C ❑ D ❑
70 A ❑ B ❑ C ❑ D ❑
71 A ❑ B ❑ C ❑ D ❑
72 A ❑ B ❑ C ❑ D ❑
73 A ❑ B ❑ C ❑ D ❑
74 A ❑ B ❑ C ❑ D ❑
75 A ❑ B ❑ C ❑ D ❑
76 A ❑ B ❑ C ❑ D ❑
77 A ❑ B ❑ C ❑ D ❑
78 A ❑ B ❑ C ❑ D ❑
79 A ❑ B ❑ C ❑ D ❑
80 A ❑ B ❑ C ❑ D ❑
81 A ❑ B ❑ C ❑ D ❑
82 A ❑ B ❑ C ❑ D ❑
83 A ❑ B ❑ C ❑ D ❑
84 A ❑ B ❑ C ❑ D ❑
85 A ❑ B ❑ C ❑ D ❑
86 A ❑ B ❑ C ❑ D ❑
87 A ❑ B ❑ C ❑ D ❑
88 A ❑ B ❑ C ❑ D ❑
89 A ❑ B ❑ C ❑ D ❑
90 A ❑ B ❑ C ❑ D ❑
91 A ❑ B ❑ C ❑ D ❑
92 A ❑ B ❑ C ❑ D ❑
93 A ❑ B ❑ C ❑ D ❑
94 A ❑ B ❑ C ❑ D ❑
95 A ❑ B ❑ C ❑ D ❑
96 A ❑ B ❑ C ❑ D ❑
97 A ❑ B ❑ C ❑ D ❑
98 A ❑ B ❑ C ❑ D ❑
99 A ❑ B ❑ C ❑ D ❑
100 A ❑ B ❑ C ❑ D ❑

CH9 Thoracotomy

Multiple Choice

1. B	11. B	21. A	31. A	41. B
2. D	12. B	22. B	32. C	42. B
3. C	13. A	23. A	33. C	43. B
4. D	14. C	24. D	34. B	44. C
5. B	15. B	25. D	35. B	45. B
6. A	16. A	26. C	36. B	46. D
7. C	17. D	27.B	37. C	47. B
8. A	18. B	28. B	38. B	48. C
9. A	19. C	29. C	39. A	49. A
10. A	20. B	30. A	40. C	50. D

True / False

51. True	61. True	71. True	81. True	91. Four
52. True	62. True	72. Vena Cava	82. True	92. Phrenic
53. 55%	63. Bicuspid	73. True	83. True	93. True
54. 91%	64. Sinoatrial	74. True	84. Ethibond	94. True
55. True	65. True	75. True	85. True	95. Two
56. True	66. True	76. True	86. True	96. True
57. True	67. True	77. True	87. True	97. True
58. Hemoglobin	68. True	78. True	88. True	98. Pleurocele
59. True	69. Arteriosus	79. True	89. Azygos	99. True
60. Pulmonary arties	70. True	80. True	90. True	100. True

CHAPTER 10

LAB VALUES CASE STUDY

LAB VALUES
Of NORMAL PATIENTS

Test	*Number*	*Capacity*
Red blood cell count		
Male	5.4 million	Per cubic millimeter
Female	4.8million	Per cubic millimeter
Platelet count	300,000	Per cubic millimeter
White blood cells	7,000	Per cubic millimeter
Eosinophils	1%	Of WBC
Basophils	1%	Of WBC
Lymphocytes	30%	Of WBC
Monocytes	6 to 8%	Of WBC
Blood PH	7.35 to 7.45	
Whole blood gases		
Capacity	17-24	vol%
Arterial		
Saturation	96-100%	Of capacity
PCO2	35-45	mm Hg
PO2	75-100	mm Hg
Bicarbonate, norm	24-28	mEq/L
Base excess	+2 to -2	(+-2 mEq/L
Venous		
Saturation	60-85%	capacity
PCO2	40-54	mmHg
PO2	20-50	mmHg
Bicarbonate, norm	22-28	mEq/L
Urine		
pH	4.5-8.0	6 is average
Specific gravity	1.005-1.030	1.015-1.024 norm
Aldosterone	2-26	Micro/ 24 hrs
Amylase	4-37	U/L 2hrs
Bilirubin and bile	Neg-0.02	Mg/dl
Calcium	7.4	Meq/24hrs
Chloride	70-250	Meq/24hrs
Magnesium	15-300	Mg/24hrs
Phosphorus	0.9-1.3	q/24hrs
Potassium	25-120	Meq/24hrs
Sodium	40-220	Meq/24hrs
Glucose	0	

5-Hydroxyindoleacetic acid	2-10	Mg/24hrs
Ketones	0	
Ammonia	30-50	Meq/24hrs

LAB VALUES
Of NORMAL PATIENTS

Creatinine clearance	100-200	ML/min
Creatinine	Male 20-26	Mg/kg/24hr
	Female14-22	Mg/kg/24hr
Protein	0-5	Mg/dl/24hr
Urea	6-17	G/24hr
Uric acid	0.25-0.75	G/24hr
Osmolality	200-1,200	MOsm/l
Porphobilinogen	0.2	Mg/24hr
Steroids		
17-Hydroxycorticosteroids	Males 5-15	Mg/24hr
	Females 3-13	Mg/24hr
17-Ketosteroids	Males 8-25	Mg/24hr
	Females 5-15	Mg/24hr
Urobilinogen	0-4	Mg/24hr
Vanillylmandelic acid	1.5-7.5	Mg/24hr
Cerebrospinal Fluid		
Cell count	0-8	Mm^3
Chloride	118-132	MEq/l
Culture	No organisms	
Glucose	40-80	Mg/dl
Pressure	75-175	Cm water
Protein	15-45	Mg/dl
Sodium	145-150	Mg/dl

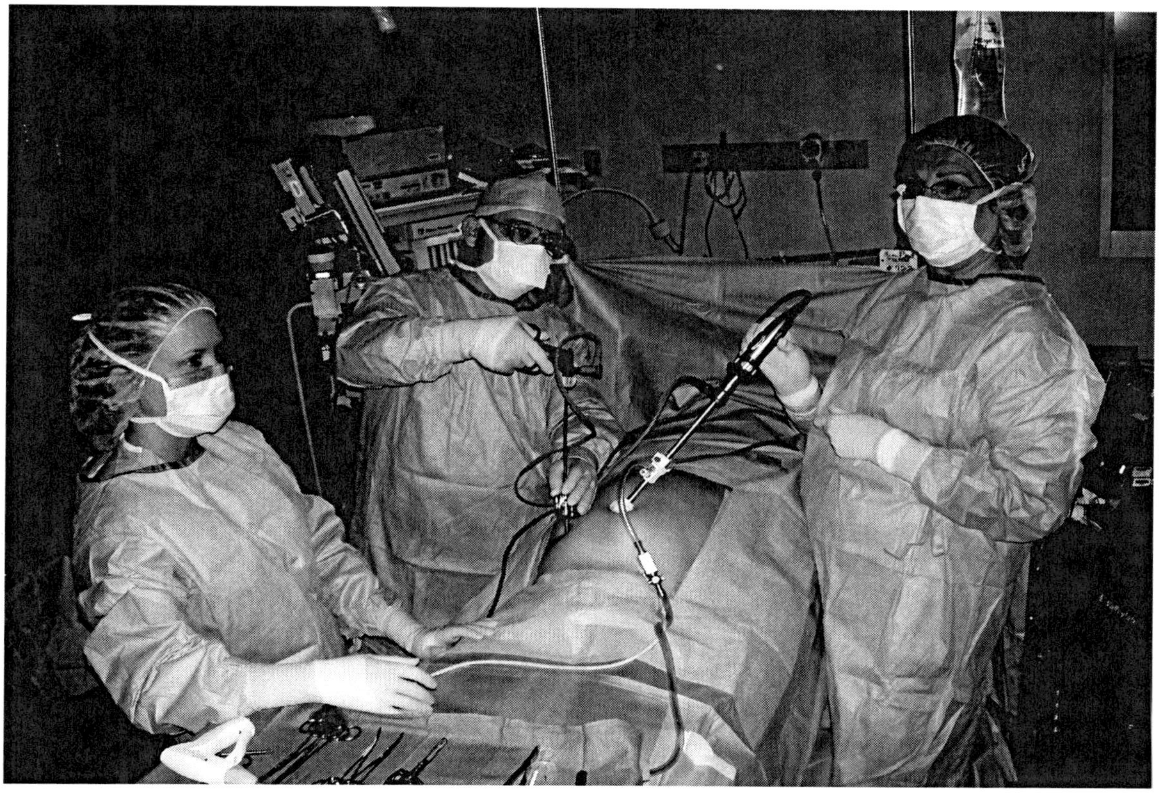

10.1

Surgical team: Dr. Don Brown, Surgical assistant Janet Robbins, Surgical Technician Kami Watson

1. Cerebrospinal culture should show.
   - A. Low organisms
   - B. High organisms
   - C. No organisms
   - D. Low counts

2. Cerebrospinal pressure should read
   - A. 75-175cm water
   - B. No water
   - C. 7.5-1.75cm water
   - D. 400cm Eq

3. Cerebrospinal cell count will read
   - A. 40mm
   - B. 80mm
   - C. 0-8mm^3
   - D. 175-200cm

4. Cerebrospinal protein will read
   - A. Low pressure
   - B. 15-45mg/dl
   - C. 100cm
   - D. 100mg/dl

5. Cerebrospinal chloride is
   - A. 118-132mEq/l
   - B. 100mEq/l
   - C. 0-8.5mm
   - D. 0-14.5mm

6. Cerebrospinal sodium should read
   - A. Low pressure
   - B. High pressure
   - C. 145-150mg/dl
   - D. 100mm^3

7. Cerebrospinal glucose should be between
   - A.0-4mm
   - B.0-8mm
   - C. 40-80-mg/dl
   - D. Non-existent

8. Blood PH should run between
   - A.4.5-5.5
   - B. 9.00-10.5
   - C. 1%-2%
   - D.7.35-7.45

9. Arterial gas saturation should be about

   A. 96-100% of capacity

   B. 45-50% of capacity

   C. 75% of capacity

   D. .50% of capacity

10. Venous gas saturation should be

   A. 60-85% of capacity

   B. 100% of capacity

   C. .5% of capacity

   D. 0%

11. Arterial PCO2 will run

   A. 100mmhg

   B. 35-45mm hg

   C. 50%

   E. 25%

12. Arterial gas bicarbonate will be

   A. 50%

   B. 2-4mm

   C. 24-28mEq/l

   D. 50mEq/l

13. Arterial gas po2 will be in what range

    A. 75-100mmHg

    B. 25-50mmHg

    C. 100-125%

    D. 0% present

14. Venous gas po2 will be in what range

    A. 0% present

    B. 75-100mmHg

    C. 50-75mmHg

    D. 20-50mmHg

15. Arterial gas base excess is normally

    A. 0% present

    B. 2%

    C. +2 to –2 mEq/l

    D. +.5 to –.7 mEq/l

16. Venous gas pco2 will be about

    A. 0.5mmhg

    B. 40-54mmhg

    C. 50-75mmhg

    D. 75-100mmhg

17. Venous gas bicarbonate norm is

    A. 100%

    B. 0% present

    C. 40-50mEq/l

    D. 22-28mEq/l

18. Red blood cell count for male per cubic millimeter should be

    A. 50,000

    B. 5.4 million

    C. 4.8 million

    D. 7,000

19. Red blood cell count for female per cubic millimeter should be

    A. 50,000

    B. 5.4 million

    C. 4.8 million

    D. 7,000

20. Whole blood cells are______ percent monocytes

    A. 6-8

    B. 10-20

    C. 50%

    D. 0%

21. Whole blood cells are ______percent Lymphocytes

    A. .75%

    B. 0%

    C. 50%

    D. 30%

22. Whole blood cells are ______percent Eosinophils

    A. 1%

    B. 0%

    C. 50%

    D. 30%

23. Whole blood cells are ______percent Basophils

    A. .75%

    B. 0%

    C. 50%

    D. 1%

24. White blood cells count run about

    A. 300 Per million

    B. 3,000 Per million

    C. 7,000 Per million

    D. 70,000 Per million

25. Platelet count in blood is

    A. 300 Per million

    B. 300,000 Per million

    C. 7,000 Per million

    D. 70,000 Per million

26. PH of normal urine is

    A.4.5-8.0

    B.8.0-10.0

    C.45

    D.7.5

27. Bilirubin and bile should read

    A. 9.0 Mg/dl

    B. 0.00

    C. Pos.-0.02 Mg/dl

    D. Neg.-0.02 Mg/dl

28. Specific gravity of urine is

    A. 1-7

    B. 1.005-1.030

    C. .105

    D. 0

29. Potassium Meq/24hrs should be

    A. 200-300

    B. -.05

    C. .005

    D. 25-120

30. Sodium Meq/24hrs should be

    A. 40-220

    B. -.05

    C. .005

    D. 25-120

31. Calcium Meq/24hrs should be

    A. 200-300

    B. 7.4

    C. .005

    D. .77

32. Chloride Meq/24hrs should be

    A. 200-300

    B. 7.4

    C. .005

    D. 70-250

33. Ammonia Meq/24hrs should be

    A.30-50

    B. 7.4

    C. .005

    D. 70-250

34. Glucose Meq/24hrs should be

    A. 200-300

    B. 0

    C. .005

    D. 50

35. Magnesium Meq/24hrs should be

    A. 15-300

    B. 7.4

    C. .005

    D. 300-400

36. Phosphorus Meq/24hrs should be

    A. 200-300

    B. 7.4

    C. 0.9-1.3

    D. 2-3

37. Ketones Meq/24hrs should be

    A. 200-300

    B. 7.4

    C. 0

    D.1-2

38. Amylase U/L2hrs should be

    A.4-37

    B. 7.4

    C. 5-6

    D. 7-8

39. Aldosterone Micro24hrs should be

    A. 2-3

    B. 2-26

    C. .50

    D. 0

40. 5-Hydroxyindoleacetic acid Mg/24hrs should be

    A. 2-10

    B. 20-30

    C. 100

    D. 0

41. Creatinine clearance ML/min should be

    A. 200-300

    B .100-200

    C. .50

    D. .50

42. Protein Mg/dl/24hrs should be

    A. 100

    B. 5-10

    C. .005

    D. 0-5

43. Urea G/24hrs should be

    A. 100

    B. 50-100

    C. .6-17

    D. 0-5

44. Uric acid G/24hrs should be

    A. 0

    B. 5-10

    C. .05

    D. 0.25-0.75

45. Urobilinogen Mg/24hr should read

   A. 0-4

   B. 5-10

   C. 100

   D. -.25

46. Male 17-Hydrocorticosteroids Mg/24hr should read

   A. 0-4

   B. 5-15

   C. 100

   D. -.25

47. Female 17-Hydrocorticosteroids Mg/24hr should read

   A. 0-4

   B. 5-15

   C. 3-13

   D. -.25

48. Male 17-Ketosteroids Mg/24hr should read

   A. 0-4

   B. 5-15

   C. 100

   D. 8-25

49. Female 17-Ketosteroids Mg/24hr should read

    A. 0-4

    B. 5-15

    C. 100

    D. 8-25

50. Vanillylmandelic acid Mg/24hr should read

    A. 1.5-7.5

    B. 5-15

    C. 100

    D. 0

51. Cerebrospinal culture should read no organisms

    A. True

    B. False

52. Cerebrospinal press should always be between 75 to 175cm water

    A. True

    B. False

53. Cerebrospinal cell count will be about

80mm

    A. True

    B. False

54. Cerebrospinal protein will read about 15 to 45 mg/dl

    A. True

    B. False

55. Cerebrospinal chloride will show 0 to 14.5mm

    A. True

    B. False

56. Cerebrospinal sodium should read low pressure

    A. True

    B. False

57. Cerebrospinal glucose should always be between 40 to 80 mg/dl

    A. True

    B. False

58. Blood ph is usually around 7.35 to 7.45

    A. True

    B. False

59. Arterial gas saturation should be about 96 to 100%

    A. True

    B. False

60. Venous gas saturation should be about 96 to 100%

    A. True

    B. False

61. Arterial PCO2 will run about 35 to 45 mmhg

    A. True

    B. False

62. Arterial gas bicarbonate will be between 24 and 28 mEq/l

    A. True

    B. False

63. Arterial gas po2 will stay in the range of 75 to 100 mmHg

    A. True

    B. False

64. Venous gas po2 will stay in the range of 75 to 100 mmHg

    A. True

    B. False

65. Arterial gas base excess is normally 20%

    A. True

    B. False

66. Venous gas pco2 will usually be about 40 to 54 mmhg

    A. True

    B. False

67. Venous gas bicarbonate norm 22 to28 mEq/l

    A. True

    B. False

68. Red blood cells count for a male are 5.4 million

    A. True

    B. False

69. Red blood cells count for a female are 5.4 million

    A. True

    B. False

70. Whole blood cells are 6 to 8% Monocytes

    A. True

    B. False

71. Whole blood cells are 6 to 8% Lymphocytes

    A. True

    B. False

72. Whole blood cells are 1% Esoinophils

    A. True

    B. False

73. Whole blood cells are 1% Basophils

    A. True

    B. False

74. White blood cell counts run about 7,000 per million

    A. True

    B. False

75. Platelet counts in blood are 300,000 per million

A. True

B. False

76. Normal urine ph is 9.45

A. True

B. False

77. Bilirubin and bile in urine should run about 0.00%

A. True

B. False

78. Specific gravity of urine is 1.005 to 1.030

A. True

B. False

79. Potassium Meq/24hr should be 25 to120

A. True

B. False

80. Sodium Meq/24hr should read 40 to 220

A. True

B. False

81.Calcium Meq/24hr should read 7.4

A. True

B. False

82. Chloride Meq/24hr should read 7.4

    A. True

    B. False

83. Ammonia Meq/24hr should read 30 to 50

    A. True

    B. False

84. Glucose Meq/24hr should read 0

    A. True

    B. False

85. Magnesium Meq/24hr should read 7.4

    A. True

    B. False

86. Phosphorus Meq/24hr should read 0.9 to 1.3

    A. True

    B. False

87. Ketones Meq/24hr should read 0

    A. True

    B. False

88. Amylase U/L2hr should be 4 to 37

    A. True

    B. False

89. Aldosterone micro 24hrs 2 to 26
    - A. True
    - B. False

90. 5-Hydroxyindoleacetic acid Mg/24hrs should be about 100
    - A. True
    - B. False

91. Creatinine clearance Ml/min should be about 0
    - A. True
    - B. False

92. Protein Mg/dl/24hrs should be 0 to 5
    - A. True
    - B. False

93. Urea G/24hrs should be about 6 to17
    - A. True
    - B. False

94. Uric acid G/24hrs should be about 0.25 to 0.75
    - A. True
    - B. false

95. Urobilinogen Mg/24hrs should read neg25
    - A. True
    - B. False

96. Male 17-Hydrocoticosteroids Mg/24hrs should be between 5 and 15
    A. True
    B. False

97. Female 17-Hydrocoticosteroids Mg/24hrs should be between 3 and 13
    A. True
    B. False

98. Male 17-Ketosteroids Mg/24hrs should read between 8 and 25
    A. True
    B. False

99. Female 17-Ketosteroids Mg/24hrs should read between 8 and 25
    A. True
    B. False

100. Vanillylmandelic acid should read between 1.5 to 7.5
    A. True
    B. False

1 A❑ B❑ C❑ D❑
2 A❑ B❑ C❑ D❑
3 A❑ B❑ C❑ D❑
4 A❑ B❑ C❑ D❑
5 A❑ B❑ C❑ D❑
6 A❑ B❑ C❑ D❑
7 A❑ B❑ C❑ D❑
8 A❑ B❑ C❑ D❑
9 A❑ B❑ C❑ D❑
10 A❑ B❑ C❑ D❑
11 A❑ B❑ C❑ D❑
12 A❑ B❑ C❑ D❑
13 A❑ B❑ C❑ D❑
14 A❑ B❑ C❑ D❑
15 A❑ B❑ C❑ D❑
16 A❑ B❑ C❑ D❑
17 A❑ B❑ C❑ D❑
18 A❑ B❑ C❑ D❑
19 A❑ B❑ C❑ D❑
20 A❑ B❑ C❑ D❑
21 A❑ B❑ C❑ D❑
22 A❑ B❑ C❑ D❑
23 A❑ B❑ C❑ D❑
24 A❑ B❑ C❑ D❑
25 A❑ B❑ C❑ D❑
26 A❑ B❑ C❑ D❑
27 A❑ B❑ C❑ D❑
28 A❑ B❑ C❑ D❑
29 A❑ B❑ C❑ D❑
30 A❑ B❑ C❑ D❑
31 A❑ B❑ C❑ D❑
32 A❑ B❑ C❑ D❑
33 A❑ B❑ C❑ D❑
34 A❑ B❑ C❑ D❑
35 A❑ B❑ C❑ D❑
36 A❑ B❑ C❑ D❑
37 A❑ B❑ C❑ D❑
38 A❑ B❑ C❑ D❑
39 A❑ B❑ C❑ D❑
40 A❑ B❑ C❑ D❑
41 A❑ B❑ C❑ D❑
42 A❑ B❑ C❑ D❑
43 A❑ B❑ C❑ D❑
44 A❑ B❑ C❑ D❑
45 A❑ B❑ C❑ D❑
46 A❑ B❑ C❑ D❑
47 A❑ B❑ C❑ D❑
48 A❑ B❑ C❑ D❑
49 A❑ B❑ C❑ D❑
50 A❑ B❑ C❑ D❑
51 A❑ B❑ C❑ D❑
52 A❑ B❑ C❑ D❑
53 A❑ B❑ C❑ D❑
54 A❑ B❑ C❑ D❑
55 A❑ B❑ C❑ D❑
56 A❑ B❑ C❑ D❑
57 A❑ B❑ C❑ D❑
58 A❑ B❑ C❑ D❑
59 A❑ B❑ C❑ D❑
60 A❑ B❑ C❑ D❑
61 A❑ B❑ C❑ D❑
62 A❑ B❑ C❑ D❑
63 A❑ B❑ C❑ D❑
64 A❑ B❑ C❑ D❑
65 A❑ B❑ C❑ D❑
66 A❑ B❑ C❑ D❑
67 A❑ B❑ C❑ D❑
68 A❑ B❑ C❑ D❑
69 A❑ B❑ C❑ D❑
70 A❑ B❑ C❑ D❑
71 A❑ B❑ C❑ D❑
72 A❑ B❑ C❑ D❑
73 A❑ B❑ C❑ D❑
74 A❑ B❑ C❑ D❑
75 A❑ B❑ C❑ D❑
76 A❑ B❑ C❑ D❑
77 A❑ B❑ C❑ D❑
78 A❑ B❑ C❑ D❑
79 A❑ B❑ C❑ D❑
80 A❑ B❑ C❑ D❑
81 A❑ B❑ C❑ D❑
82 A❑ B❑ C❑ D❑
83 A❑ B❑ C❑ D❑
84 A❑ B❑ C❑ D❑
85 A❑ B❑ C❑ D❑
86 A❑ B❑ C❑ D❑
87 A❑ B❑ C❑ D❑
88 A❑ B❑ C❑ D❑
89 A❑ B❑ C❑ D❑
90 A❑ B❑ C❑ D❑
91 A❑ B❑ C❑ D❑
92 A❑ B❑ C❑ D❑
93 A❑ B❑ C❑ D❑
94 A❑ B❑ C❑ D❑
95 A❑ B❑ C❑ D❑
96 A❑ B❑ C❑ D❑
97 A❑ B❑ C❑ D❑
98 A❑ B❑ C❑ D❑
99 A❑ B❑ C❑ D❑
100 A❑ B❑ C❑ D❑

CH10 Lab Values

Multiple Choice

1. C	11. B	21. D	31. B	41. B
2. A	12. C	22. A	32. D	42. D
3. C	13. A	23. D	33. A	43. C
4. B	14. D	24. C	34. B	44. D
5. A	15. C	25. B	35. A	45. A
6. C	16. B	26. A	36. C	46. B
7. C	17. D	27. D	37. C	47. C
8. D	18. B	28. B	38. A	48. D
9. A	19. C	29. D	39. B	49. B
10. A	20. A	30. A	40. A	50. A

True / False

51. True	61. True	71. 30%	81. True	91. 100 to 200
52. True	62. True	72. True	82. 70 to 250	92. True
53. 0-8mm^3	63. True	73. True	83. True	93. True
54. True	64. 20 to 50	74. True	84. True	94. True
55. 118-132mEq/l	65.+2 to-2	75. True	85. 15 to 300	95. 0 to 4
56 .145-150mg/dl	66. True	76. 4.5 to 8.0	86. True	96. True
57. True	67. True	77. Neg-0.02	87. True	97. True
58. True	68. True	78. True	88. True	98. True
59. True	69. 4.8 Million	79. True	89. True	99. 5 to 15
60. 60 to 85%	70. True	80. True	90. 2 to 10	100. True

CPSIA information can be obtained at www.ICGtesting.com
Printed in the USA
LVOW092102210213

321070LV00001B/108/P

9 780615 264899